Luminos is the Open Access monograph publishing program
from UC Press. Luminos provides a framework for preserving and
reinvigorating monograph publishing for the future and increases
the reach and visibility of important scholarly work. Titles published
in the UC Press Luminos model are published with the same high
standards for selection, peer review, production, and marketing as
those in our traditional program. www.luminosoa.org

Governable Spaces

Governable Spaces

Democratic Design for Online Life

———

Nathan Schneider

Illustrations by Darija Medić

UNIVERSITY OF CALIFORNIA PRESS

University of California Press
Oakland, California

Suggested citation: Schneider, N. *Governable Spaces: Democratic Design for Online Life*. Oakland: University of California Press, 2024. DOI: https://doi.org/10.1525/luminos.181

This book is freely available in an open access edition thanks to TOME (Toward an Open Monograph Ecosystem)—a collaboration of the Association of American Universities, the Association of University Presses, and the Association of Research Libraries—and the generous support of the University of Colorado Boulder Libraries. Learn more at the TOME website, available at: openmonographs.org.

Library of Congress Cataloging-in-Publication Data

Names: Schneider, Nathan, 1984- author. | Medić, Darija, illustrator.
Title: Governable spaces : democratic design for online life / Nathan Schneider; illustrations by Darija Medić.
Description: Oakland : University of California Press, 2024. | Includes bibliographical references and index.
Identifiers: LCCN 2023036722 (print) | LCCN 2023036723 (ebook) | ISBN 9780520393943 (paperback) | ISBN 9780520393950 (ebook)
Subjects: LCSH: Internet governance. | Online social networks—Political aspects. | Democracy. | Feudalism—Political aspects. | Social media and society.
Classification: LCC TK5105.8854 .S36 2024 (print) | LCC TK5105.8854 (ebook) | DDC 384.3/34—dc23/eng/20230908

LC record available at https://lccn.loc.gov/2023036722
LC ebook record available at https://lccn.loc.gov/2023036723

33 32 31 30 29 28 27 26 25 24
10 9 8 7 6 5 4 3 2 1

CONTENTS

Introduction

Democracy in the Wild

Imagine a gathering under a tree, a couple dozen people sharing a picnic in a park. The day begins clear, good for cooking and playing and lying on blankets. Food and games are out, splayed around the tree and the lawn around it. As the afternoon goes on, clouds form and gather overhead, but few of the picnickers notice until the first raindrops fall. Murmurs begin to spread, bodies agitate. The murmurs all amount to some version of the same question: *What should we do?*

A choreography of rough consensus is underway. The networks of friends at the picnic activate, checking in with each other using words and how they carry their bodies. Some hold themselves high, determined to wait out the weather, while others look around skittishly, assessing the quantity of rain and the perceptions of others. Friends cross-pollinate information across the clusters of family. Within families, members seem to look toward one or two of them—an elder who speaks only the old language or a volatile kid or a guest, depending on the family—to make the call that the rest will follow. A ranger from the park service comes by, an agent of the regional government, to offer a warning about the perils of being under a tree during a thunderstorm.

The air begins to smell of petrichor as moisture fills the pores of stones and dirt, releasing as aerosols the oils they have been holding inside them. By then, most of the birds and squirrels nearby already know what is coming from the changing barometric pressure, and they are back in their nests. The tree alters the chemicals oozing from its roots, which the mycelial networks underneath transmit across that section of the park. Worms weaving among them feel the moisture and move upward toward the surface, into the rain that others are trying to escape.

Enough families leave that even the picnickers most determined to stay no longer see the point. The thick air and rush of creatures have enveloped what is left of the human activity. Those remaining people now seem isolated and wandering, no longer cohering as a single event like they had just a few minutes earlier. The critical mass that made the place a picnic had gone.

By then word has spread about a group chat. There, they can share photos and find their lost things that others might have hastily gathered up. What was before, at the picnic, an uneven topology of social location and circumstance now becomes an instantaneous ledger of opinion. One phone after another logs in there, lighting up with chatter about whether the picnic should have ended. But this time the youngest people do not have the equipment to add their voices; the eldest tend to have trouble joining. Lightning never came, and before long the rain is gone.

Go back?
We came all that way to get there!
Nah, already packed up.

In the chat, everyone is a speech bubble. There are some side chats among friends, but the main group flattens the textured structures of relationship. Disagreements fly by, but nobody is sure what would be the criteria for a decision or how to signal commitment. The chatter ricochets back and forth. Some who were quiet under the tree feel more free to speak up here. One person complains especially crudely, only to vanish from the chat—removed by the person who started it, whom the software regards as its admin. Factions form and dig in their positions. Notifications announcing messages continue to flash on the remaining people's phones, until the futility of the debate slows them to an occasional emoji, and then some photos taken earlier, and then no more.

What happened to the picnic when it went online? This is a version of the questions many of us find ourselves asking over and over, as one scene of social life after another migrates to digital networks—our workplaces and markets, our classes and clubs, our money and family, our religion and politics. The answers, as above, are never straightforward. But they are increasingly consequential.

This is a book about the politics of everyday life, and everyday online life in particular—among the internet-borne social spaces where people see each other and interact through digital tools. I contend that the most quotidian kinds of online politics, such as those in the tale above, affect the flows of power at the largest scales. The ways people can and cannot collectively self-govern in daily online life, furthermore, have been constrained in dominant social networks. I will argue that the constraints on governance in online spaces have contributed to the peril of democratic politics in general. It is not enough to merely defend existing governmental institutions; healthy democracy depends on enabling creative new forms of self-governance, especially on networks.

Several proposals flow from those claims. One is the need for online communities themselves to self-consciously cultivate democratic practices. These practices can serve as the basis for a social-media design paradigm that invites diverse kinds of community governance to emerge and flourish. But community-scale democracy will remain only marginal within antidemocratic infrastructures. A further paradigm is therefore necessary for the policies encoded in law and technical systems that organize online life—self-governance, rather than top-down authority, as the basis for problem-solving. Such a paradigm would make networks home to new jurisdictions—enabled by but not always reducible to the jurisdictions of geographical territories.

Much of this book dwells in interactions of human politics and technological systems. But, as above, the more-than-human world envelops it all, providing the stage and the stakes: a planet waiting to see whether we can govern our way out of self-destruction, deciding whether to maintain the conditions necessary for human civilization.

Is there democracy in the wild?[1] Creatures hurtling through space on a fragile world can expect no rights or powers of decision from physics and biology. A government's claim to rule means little in a high-mountain wilderness or in a neighborhood whose residents have made themselves ungovernable to survive against a hostile police force. Yet *governance* and its cognates are names we use for doing what all life-forms must: orchestrating our perceptions and reactions so as to have a chance at thriving in our surroundings. Consider it simply the intersection of power and cooperation—an intersection hardly unique to us.[2]

Any precise meaning of *self-governance* is necessarily contextual, depending on who is involved and what kinds of say they seek. Likewise, I claim no fixed definition for *democracy*. I understand it as always a horizon, a longing for power shared equitably among participants, a destination that moves depending on where one stands.[3] An orchestra permits hierarchies intolerable to a punk band, but the people in each may still see themselves as living toward democracy. If democracy is the horizon, self-governance is a plausible practice for moving in that direction. *Governable spaces*, then, are where democratic self-governance can happen.

The story of the picnic included different kinds of spaces and, among them, missed opportunities. What if other picnickers had heard those only comfortable speaking up online? What if the group chat had included tools for steering debate into decision? What if the picnickers had been more skilled at making decisions online because they were used to having and using real power?

The online networks that are the subject of this book are a kind of wilderness. They are evolving biomes, host to a polyphony of people and machines. The networks are not fully apart from the governments that claim to rule the world, but not entirely subject to them either. What happens online is terrible and wonderful;

I love my favorite online haunts. If I criticize our networks as they are, it is because I see glimpses of the governable spaces they could become. Our networks are spaces we have still only begun to co-create and self-govern and thus to make our own.

DEMOCRATIC EROSION

It is by now a truism that democracy is in decline around the world. Political scientists have diagnosed the "erosion" or "deconsolidation" of democratic institutions among governments, as well as in global opinion polls, which exhibit collapsing affection for democratic ideals.[4] Countries such as the United States, the world's longest-running constitutional democracy, and India, the world's largest, have voted into power regimes with autocratic tendencies. Other countries of diverse kinds, from Hungary to the Philippines, have both led and followed. According to one analysis, between 2011 and 2021, "toxic polarization" dividing political factions spread from five countries to thirty-two; the number of countries with worsening freedom of expression went from five to thirty-five; and the share of the world's population living in autocracies increased from 49 percent to 70 percent.[5] The situation means trouble for those who regard democratic government as an intrinsic good, to be sure. It also bears other dangers, threatening a self-reinforcing spiral of authoritarianism, economic exploitation, and environmental destruction, especially as leaders seem to regard protecting ecological and social health as an unacceptable constraint on their mandates to achieve national greatness.[6]

Blame for democratic erosion falls in many directions, from intersecting inequalities and climate-induced migration to widespread corruption and insufficiently civic-minded elites.[7] But it is hard to avoid laying blame on the absorbing, distracting, glowing presence that has reconfigured public and private life for so many of us in recent decades: online social media. Scholars and journalists have argued that social networks have worsened polarization, provided mouthpieces for authoritarians, enabled violent extremists to organize, and undermined trust in institutions.[8] Additionally, mounting evidence suggests that users perceive online platforms themselves as unaccountable polities, resulting from experiences of arbitrary rule enforcement, a lack of due process, and an absence of sensitivity to context.[9] The diagnoses, in turn, produce calls for a response. Proposals typically take the form of fresh impositions of consolidated power, whether through governmental regulation of platform companies, takeovers by billionaires aspiring to be saviors, or the fiat of platform companies themselves.[10]

Meanwhile, social-media-savvy protest movements have set out to reinvent democracy with viral mobilizations, denouncing old regimes and experimenting with self-governance in the streets. The year 2011 saw a wave of uprisings spread from the Middle East, across Europe, to Wall Street, and then around the world again. Protesters often eschewed representative democracy and modeled forms more responsive, creative, and direct. But in the years since, hardly any gains from that period have stuck, and in most cases the authoritarians have only tightened their grip. Civil wars with their roots in those protests—in Libya, Syria, and Yemen—are still smoldering. Movements have succeeded in using online tools to spread their messages and cause fleeting disruptions, but those achievements have not translated into lasting democratic blocs that have shifted power in meaningful ways.[11]

Even if the Internet is neither a complete nor satisfying explanation for eroding democratic norms, there is reason enough to believe that aspects of networked life have contributed to aspects of democratic erosion. The growing ubiquity of online networks seems to have roughly preceded the rise of the new aspiring dictators. Those figures, more than trying to restrict and censor social networks, have embraced them as their own. Social algorithms often privilege the kinds of polarizing, abusive messages that undermine civil discourse. And rising levels of app-fueled anxiety might leave people more susceptible to promises of autocratic certainty.

This book will add one more accusation to the pile: the design of online social spaces has contributed to the atrophy of everyday democratic skills. The diagnosis also bears remedies. More than other explanations of democratic erosion, this account suggests that the future of democracy can begin at the level of ordinary community, wherever we find ourselves together, where each of us has the chance to make a difference.

EVERYDAY DEMOCRACY

To measure the situation of the digital, consider the analog. While I was beginning the research that led to this book, I was receiving regular updates from my mother on her neighborhood garden club. The club has survived from the heyday of suburban housewives—which my mother, as a retired government employee, never was. But the club elected her president. She described to me the debates, the subtexts, the meetings, and her stratagems for facilitating the process.

The club's bylaws occupy eight pages in an annually printed, thirty-eight-page handbook. It also has chapters on hospitality and flower arranging. The bylaws' structure includes articles, sections, and enumerated subsections. As a legal document governing a nonprofit organization, the language is formal, with lots of "shall" statements and capitalized terms. The club members don't normally talk this way with each other. But when they have decisions to make or conflicts among them, they can flip to those pages and find a path forward. The bylaws help make the club a governable space.

As she talked about the club, my mind drifted to my own recent encounters with governance: running a five-hundred-person email discussion group, lurking among open-source software communities, and documenting hashtag protest movements. As an admin in online spaces, I struggled with how to adopt basic democratic practices like those of the garden club. The interfaces I had to navigate in those spaces provided no guidance. There was no functionality for elections, no

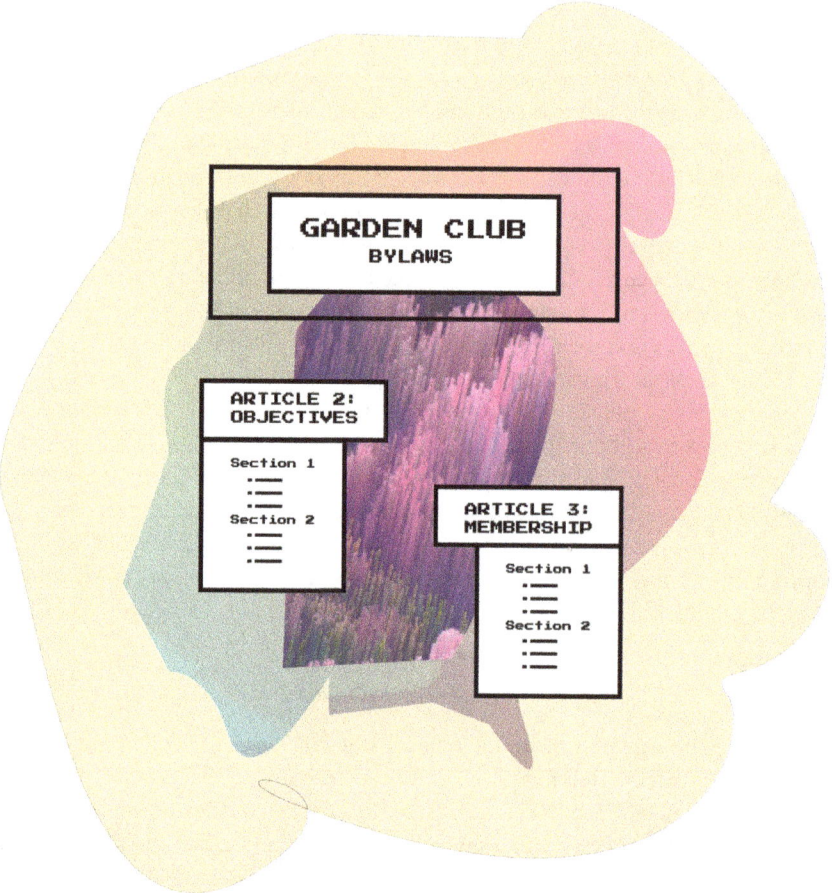

FIGURE 1.

mechanisms for dispute resolution, no template for simple bylaws. I could patch together a vote or summon a jury on my own, sure, but what would count as a decision? On what basis could I establish ground rules, and what if I didn't want to implement the outcome? Ultimately, power rested with me and whoever else's accounts had admin privileges. What would it mean for other users to hold us admins accountable? Few online groups I had been part of could hold a candle to the simple and effective set of rules that had governed the garden club since the 1960s, rules unremarkable among countless similar organizations with a vast range of purposes. Few online groups will last so long.

My mother's garden club inherits a legacy of second-nature civic association that impressed the French aristocrat Alexis de Tocqueville when he toured the United States in 1831. In contrast to late-monarchical Europe at the time, he was taken with how fervently Americans seemed to form organizations, for all kinds of interests and purposes. It struck him that the lessons learned in community-scale groupings had something to do with the practice of the representative government, still nearly unique to the United States at the time: "The greater is the multiplicity of small affairs, the more do men, even without knowing it, acquire facility in prosecuting great undertakings in common. Civil associations, therefore, facilitate political association: but, on the other hand, political association singularly strengthens and improves associations for civil purposes."[12]

Meanwhile, he surmised that when people do not have experience in self-governing associations, they fear the risks of it and doubt their capacity to participate. Democratic muscles need exercise: "When [people] are as yet but little versed in the art of association, and are unacquainted with its principal rules, they are afraid, when first they combine in this manner, of buying their experience dear. They therefore prefer depriving themselves of a powerful instrument of success to running the risks which attend the use of it."

Tocqueville anticipated thinkers such as John Dewey and Paulo Freire in articulating the interrelation of politics and education. Democratic society works only if people are educated for it, and education cannot be democratic without involving direct political engagement. "Political associations may . . . be considered as large free schools, where all the members of the community go to learn the general theory of association."

Tocqueville wrote passages like these with particular sensitivity to the anxieties of his fellow European elites, who were in the habit of suppressing popular associations for the sake of social stability. For the aristocrats' benefit, Tocqueville took particular pains to explain how widespread association would actually serve the social order rather than undermine it. The more invested people are in their own endeavors, he argued, the more stake they have in the order on which it rests: "[If you] perceive that the Americans are on every side unceasingly engaged in the execution of important and difficult plans, which the slightest revolution would throw into confusion, you will readily comprehend why people so well employed are by no means tempted to perturb the State, nor to destroy that public tranquillity by which they all profit."

Perhaps the same is true of online mobs, scammers, and trolls. Would they too have less incentive to disrupt if they had more stake, if they had their own mini-democracies to care for?

The bylaws of the garden club and the associations Tocqueville admired would not translate straightforwardly online. Too much is different in online spaces: the ease of joining and leaving, the cultural and geographic diversity, the speed, the anonymity, the metrics of reputation, and on and on. And yet his basic insight has remained salient: a synchrony binds the smaller and larger scales of political life. Findings that correlate democratic government and everyday civic associations persist long after Tocqueville's time, across diverse contexts.[13] Causal "spillover effects" indicate that when people participate in local democratic activities, they are more likely to involve themselves in the affairs of government.[14] Among social movements, practicing democracy at small scales has often been a strategy for building democratic power at the highest levels. For instance, the modern cooperative movement first took hold in England among Chartists, factory workers demanding the right to vote in elections. To exercise and prove their democratic skills, they formed cooperative stores where every customer had a vote. The English cooperators became allies to US slavery abolitionists like Frederick Douglass, and cooperatives in turn became important features of Black liberation movements from civil rights to Black Lives Matter. Nineteenth-century Populist organizers in the American West saw local cooperatives and other associations as the best defense against the appeal of demagogues to exploited farmers.[15] More recently, sociologist Erik Olin Wright understood participatory associations as "real utopias" that contribute to a social change through "interstitial transformation."[16] These have been the offline governable spaces that help make democratic politics possible.

The political significance of ordinary life need not stem from activities that are distinctly civic or economic. What about walking to the train station, watering a community garden, or teaching a child to repair a toy? I draw also from theorists of everyday life since Tocqueville who have found politics in the kinds of activities that seem farthest from it, that dominant cultures render as officially insignificant.[17] Michel de Certeau and Henri Lefebvre identified the everyday with tasks of domesticity and social care; the everyday I focus on looks more like busy fingers and eyes tracking screens or moving through a physical world while preoccupied with what took place on a server elsewhere. In such moments lie opportunities for critique and meaning-making, resistance and world-building. I follow Anne Norton's insistence that "sovereignty is a commonplace" held in our bodies and communities, not an "exception" from above as prominent political theorists have claimed.[18] Before twentieth-century feminists said it better, Tocqueville taught political thinkers to notice that the personal, especially the interpersonal, is political.

Tocqueville's perceptions, however, lead to places I cannot follow. He failed to see the genuinely democratic possibilities among people facing European colonization from Africa to the Americas—advocating a crusade of democracy through conquest rather than against it.[19] For this reason and more, in these pages I rely on another a lineage of political thought, which took as its starting point anticolonialism and anticapitalism, then expanded later into ecological feminism. The lineage begins with the Trinidadian writer C. L. R. James, then passes to the Chinese-American organizer and philosopher Grace Lee Boggs, James's longtime collaborator, and then to adrienne maree brown, a disciple of Boggs in Detroit who has become a pivotal voice in present-day activism surrounding climate justice, Black liberation, queer identities, and science fiction. James, Boggs, and brown share with Tocqueville that critical concern about how the texture of the everyday might contribute to the transformation of the world, but they see openings that he did not.

James wrote a definitive history of Haiti's independence struggle, *The Black Jacobins*, and he played a guiding role in decolonizing Africa. Among his writings is a short essay from the mid-1950s, "Every Cook Can Govern," which imagines labor unions reviving ancient Greek direct democracy by appointing officeholders at random from the community.[20] What would our politics look like, he asks, if we really believed that each of us has the right and ability to self-govern? What kinds of people could we cultivate if we held that trust in each other?

These are questions Boggs explored deeply in the context of labor organizing among Detroit factory workers. Later in her life, after parting ways with James, she organized a youth summer camp, became fascinated with new decentralized technologies, and studied systems of self-organizing in biology. She mentored several generations of activists, teaching them to ask questions and hold faith in people to discover their own answers when given the chance.[21] And brown has continued those explorations through her practice as a social-movement facilitator and writer, grounding the work of struggle and social change in the experience of friends in a group chat, in bodily pleasures, in theories about fungi and fractals.[22] She notices how communities, like fungi, build subterranean connections through networks; like fractals, people's ordinary interactions with loved ones and neighbors shape the possibilities of politics at the largest scales. The faith in people's capacity to self-govern that animated James's anticolonialism and Boggs's devotion to the possibilities for Detroit becomes, for brown, an antidote to the mayhem of very-online life, helping her douse such flame wars as "cancel culture" and the backlash to "defund the police." Together, James, Boggs, and brown see transformative power in even intimate governable spaces.

These three are not usually considered media scholars, although I have learned a lot by reading them that way. Throughout this book I draw them into a shared conversation about making an inclusive, accountable, networked democracy. I do so not to detract from the urgency and centrality of any specific struggle. Building governable online spaces could enable more powerful, creative movements, but I do not mean to prioritize that strategy over others. I hope to invite a conversation that follows Aníbal Quijano's understanding of "totality," a search for holistic, cross-cultural knowledge that welcomes difference and refuses domination.[23] The crisis of self-governance is in many respects a shared crisis around the world, even as it appears to us through many different histories, experiences, and disguises. The rot seeps everywhere, but it does not everywhere smell alike.

Life can flourish on rotting logs, as brown's fungi remind us. If nation-state democracy is rotting, then we might allow ourselves to imagine its erosion not solely as a loss. Rot is metabolism, an act of digestion into something else. If democracy is not a static organism so much as an evolving symbiosis, then we can allow ourselves to search for more of the possible feedback loops that we could sense and act on.[24] The subject at hand is sensual, even while it is a matter of technology.

ARTIFACTS AND POLITICS

There is no more notorious error in the study of media technologies than determinism—interpreting some device as single-handedly steering social outcomes and thereby denying the role of people in shaping their own cultures and power structures. I admit at the outset to edging around that theoretical sinkhole. This book rests on a claim that the dominant design patterns of social-media technologies have constrained social and political possibilities, including the cultural options and possible power structures. Democratic self-governance is far harder than it needs to be in online spaces, and autocratic flows of power arise easily—not so much because of the people as because of the tools and the economies that reinforce them. Different tool designs can make self-governance easier to practice and improve. To borrow the canonical phrasing of Langdon Winner, who tangled with determinism too, these artifacts have politics.[25]

Tarleton Gillespie ends his field-defining book on platform governance, *Custodians of the Internet*, with a proposal that ordinary users should have greater involvement in the rule of online space and that platform companies must "share the tools to govern collectively."[26] Probing that proposal and then attempting to make good on it turn out to be far easier said than done. Technological inertia, combined with allied forces in business models and culture, has produced counter-democratic tools. Collective governance runs contrary to how online spaces have typically taught us to behave in them. Gillespie's proposal therefore requires amending. To "share the tools" as the tools are will do little for governing collectively. The tools themselves must be different for governable spaces to emerge.

That is where I slip out of deterministic trouble. It is through the practice of intentional self-governing that people can begin rethinking and remaking their tools. Tools constrain politics, but people can fashion better tools with politics and business models that do not take corporate control as the starting point. I will follow, for instance, Philip E. Agre's call, at the enigmatic end of his career as an engineer and humanist, for the cultivation of "political skills." Agre stressed that a healthier politics should begin and end with human practices, even while rethinking the technologies in between. The task is well captured in Ruha Benjamin's inversion of an old Facebook slogan: "Move slower and empower people."[27] As in the Slow Food movement, *slow* is less a matter of velocity than of making time to observe and attend to the relationships at play.

Andreas Hepp's formulation of "deep mediatization" points a further way out of determinism. Under this condition, Hepp writes, "all elements of our social world are intricately related to digital media and their underlying infrastructures."[28] If society has become so thoroughly mediated, how could we expect democracy to emerge in not-especially-democratic media? Hepp shows how algorithms and data aggregation do not just communicate but reshape society. Of similar importance, I argue, are the interfaces and administrative features of online social spaces, the sites that manifest who has power over whom. These user experiences organize what Hepp identifies as the "figurations" of mediated life: the complexes of institutions and their participants engaged in "embodied doing." Governance occurs through figurations, too. The later chapters of this book move toward refiguration, or reorganizing certain figurations in more democratic directions. I attempt to set in motion a sequence of what Hepp calls "recursive transformation."[29] This involves not a single intervention but interventions across mediated life. With alternating social, technical, and economic proposals, I outline a cyclical theory of change, turning from multiple directions.

As the argument progresses, it should become clear that technical solutions alone are inadequate—and impossible—even for problems that people experience most directly through technical interfaces. Those interfaces come to us not by their own accord but through the deployments of capital and power that orchestrate their design.

I will not stop at political economy, however. Social structures and media systems depend on the life-forms that create them, the biological and creative forces that call into question any attempt to take systematizing too far. I follow Sarah Kember and Joanna Zylinska's *Life after New Media* in their emphasis on *life*. They cast media studies as constituting a "theory of life," involving "the interlocking of technical and biological processes of mediation." In these terms, we can allow ourselves to think about fungi as media, to take seriously the habits and rituals involved in making an online place feel like home. Mediation constitutes a cyborg organism. On that assumption, we can more fully exit the dichotomy of user and machine, of determiner and determined. The possibility of self-governance rests on recursion, again, between biology and technology, the self and the network, the creative and the critical. Kember and Zylinska introduce themselves as artists as well as scholars, modeling an interplay of analysis and intervention—a "creative mediation" that they summarize as simply "doing media studies."[30] Doing-through-study is what I aspire to here.

I have been aided in that doing by being holder of a key to the Media Archaeology Lab, located in a basement half a block from my office at the University of Colorado Boulder.[31] The lab houses multitudes of functioning and supposedly obsolete computers, games, mobile devices, and technical manuals, available for use in study and artist residencies. This feat of maintenance has reminded me to test my ideas in living relationship with machines, playing with them and relying on them. Media archaeology serves as a helpful frame for the orientation to history here: the past is of interest mainly to the extent that it still lurks among us in the present, including those parts of the present that declare themselves as innovation. But the Media Archaeology Lab is no mere curiosity shop; in my sessions there I work alongside artists and hackers composing new works with the machines that have survived from past product cycles. From the past, they carry possible futures. The real usefulness in seeing the world as mediation is the extent to which it becomes an invitation for recasting molds of meaning in software code, for performing social experiences that code could never capture.

DEMOCRACY AS A DESIGN PRACTICE

Zizi Papacharissi has recently wondered, "What if democracy is not what we are after but the path to something else?"[32] It is a question the eminent communication scholar posed not only to herself and her readers but to one hundred interview subjects around the world. In many of those conversations, her informants did not seem to have the words to describe either the problem or the path forward. They could agree only on the sham in their governments' claims to be democracies. Nobody expressed enthusiasm for the people representing them. "We have turned democracy into a rigid routine," Papacharissi concludes.[33]

Perhaps leaning so hard as I have on *democracy* will only cause it to snap. Perhaps we need another word; perhaps the word can be refurbished and put to better use. Either way, technology is sure to be drafted in the cause. A further fruit of Langdon Winner's reflections on artifacts and politics is an observation about the amnesia that surrounds incidents of innovation: "In our times people are often willing to make drastic changes in the way they live to accord with technological innovation at the same time they would resist similar kinds of changes justified on political grounds."[34]

Technologies can open political doors that ordinary politics may not open alone. We see this pattern in governments' willingness to let ridesharing apps categorically violate labor law or for nuclear weapons to justify consolidating the authority of a chief executive.[35] That's the danger in determinism: the excuse that technology left no other choice. But in a world where the range of political possibilities can seem close to nil, this amnesia in the face of gizmos occasions a weird and perhaps necessary hope.

I contend several technological ruptures are underway that all present opportunities for democracy or whatever the future needs to call it. These ruptures represent contested spaces, not salvific solutions. They present as many dangers to democratic politics as opportunities, and *how* they proceed matters at least as much as *whether*.

One rupture involves initiatives among territorial governments that introduce forms of citizen voice, often with new media in hand. These range from the advent of participatory budgeting processes in Porto Alegre, Brazil, in 1989 to the digital deliberation platforms adopted more recently in places like the city of Barcelona and the national government of Taiwan. The experiments include wiki-style efforts to crowdsource constitutions, assemblies of randomly selected citizens drafting policy proposals, and the use of artificial intelligence to identify clusters of participant opinion independent of political parties. Even under Chinese authoritarianism, such forms of consultation have flourished. Efforts to institutionalize restorative justice or practice transformative justice prefigure societies less reliant on police and incarceration. In certain times and places there seems to be at least partial openness among governments to explore more information-rich feedback loops than periodic elections. But in most cases the innovations perform merely advisory roles, granting citizens little in the way of new powers that are meaningfully binding. As such, these forays also disclose the resistance of today's territorial governments to departing from what Papacharissi calls their "rigid routine."[36]

Another rupture is the advent of what goes by the names of blockchain, Web3, or simply crypto—the circus of innovations and crises that have arisen since the release of the Bitcoin cryptocurrency in 2009. Crypto-based communities, organizations, and protocols have implemented novel decision-making procedures and organizational structures on and off the immutable ledgers of their blockchains. The reliance on open-source software means that when something works, it can spread rapidly to other communities. Regardless of any failures to fulfill what advocates have promised for it, I argue that this rupture is important because of the almost surgical precision with which crypto's distributed ledgers differ in their power structures from earlier online systems hosted on central servers. Much in the realm of crypto is decidedly antidemocratic and unabashedly plutocratic, but its rise—and even the appalling hype of its speculative cycles—presents an opportunity for reimagining networks along more democratic lines.[37]

The quest for governable spaces is a chance to design. Democratic design does not come easily to many of us, however. Too often we regard democracy as either a condition fixed long ago in a constitution or indefinitely out of reach, depending on how we experience the governments under which we live. But to design digital spaces as governable spaces means that we might have the chance to define and redefine democratic practice far more frequently than the drafting of a constitution every few centuries. Designing the media of governance on social networks, for instance, could become as valuable a skill as jockeying for power.

My approach to design owes homage to several sources. One is Arturo Escobar's framework of "designs for the pluriverse," which insists that no single design can serve all people and cultures and that we should regard design as an exercise in historical consciousness and multiplicity. Escobar also sees design through a decolonizing lens, as a form of resistance to being designed from elsewhere. The framework of "design justice" further insists that design must occur through rigorous accountability to the people whose lives it will shape; it emerged out of the Allied Media Projects network in Detroit, among disciples of Grace Lee Boggs, and has been crystallized in the work of Sasha Costanza-Chock.[38] Part of what governable spaces must enable is the ability to craft and practice that accountability.

Another approach to designing deeper accountability derives from the cybernetic school, which views human, ecological, and technical systems through the structures of their information flows and feedback loops. Salvador Allende's attempt to create a governable computer system in Chile, Project Cybersyn, sought to organize these flows at the scale of a country. I draw also from scholarship on mechanism design and common-pool resources, particularly in the vein of Elinor Ostrom, a literature that complements democratic ideals with insights from economics and game theory. Finally, with Tocqueville, I regard democratic design as, in important respects, a matter of spiritual imagination, a mediation between transcendent aims and immanent conditions.[39] The invitation to design comes with many more invitations wrapped within it.

Together, these lines of thinking stress that design does not occur in a vacuum or in the head of a solitary designer. It emerges through social and economic life, which shapes and constrains it. To change how we design means also changing aspects of the social order. Enabling democratic design in online life, I will argue, will involve redirecting the flows of finance and regulation. To change these flows is to alter the conditions of design. I think we can build what Ivan Illich called "tools for conviviality"—tools that support "autonomous and creative intercourse among persons, and the intercourse of persons with their environment." Convivial tools are ones that invite us to be creative and responsible, rather than deferring responsibility to someone else. Illich warns, however, that achieving conviviality is possible "only if we learn to invert the present deep structure of tools."[40]

I should acknowledge some contexts of my own design and the design of this book. I have thought and written in ongoing conversation with hundreds of collaborators in the Metagovernance Project, a community of research and practice that I have had the opportunity to help lead.[41] Through Metagov, I have found co-authors, co-investigators, co-developers, and co-critics, all of whom share a commitment to advancing the possibilities of self-governance in online spaces. One way of phrasing the purpose of this book is to argue for the value of what they are all up to—what we are up to together.

I have also come to see the need to acknowledge the sources of my own at-times outsized faith that human beings are capable of democracy in the first place. There are several. My participation in the tradition of Catholic social teaching, for instance, has taught me to regard self-governance at proper scales as a right and obligation of human dignity. I am moved, for instance, by the deceptively modest aspiration of the Catholic Worker movement to form "a world where it is easier to be good." For much of the past decade, also, I have worked closely with and learned from the founders of a new generation of cooperative businesses, practicing economic democracy in the tech industry and elsewhere.[42] But the experiences that come to mind most frequently occurred at the school I attended as a teenager, a public high school whose founders insisted on making it unusually democratic. There I took part in setting the school's rules at the weekly "town meetings" and had the opportunity to lead the design of a new admission policy after a court struck down an earlier one. Knowing well my own lack of formal preparation for these tasks, I became convinced that if people are given a real chance to self-govern, with the guidance and infrastructures they need to do so, they will rise to the occasion. The years since have left me less optimistic that self-governance is a clean and easy answer to any question, but my hope in people's ability to surprise themselves with it remains.

The persistence of those early experiences for me, decades later, testifies to the power of designing governance experiences. When people participate in healthy democracy firsthand, it can leave a lifelong impression that such a thing is possible, even if actual manifestations of it remain rare. Those experiences are why Tocqueville's associations and brown's fractals ring so true to me. To design the governance of even minute comings-together is to shape what people feel they are capable of. Architects, lawyers, and decorators have long ordered public spaces for self-governance through their designs. The same thing can happen in the design of governable spaces online.

HORIZONS AND LIMITATIONS

The chapters that follow undertake a journey from an archaeology of pre-internet software to a call for rethinking the governance of global networks. In the process, I will argue for reorienting habits around online spaces from deskilling to political skills, from server control to community control, from paternalism to governability. Toward that end, I offer a sequence of concepts that constitute a vocabulary for online democracy.

I begin with a diagnosis of *implicit feudalism*, the dominant design pattern for online spaces, in which all power derives from founders and admins, and most users lack opportunities for direct, instrumental *effective voice*. The second chapter makes a case for the far-reaching consequences of this kind of design and its affinity with the ideology of *homesteading*, which extends the trajectory of American colonization into the digital economy. There, I contend that the structure of daily online life has prefigured the rise of authoritarian urges at the level of national governments. Democratic erosion coincides with shortage of democratic practice when social life migrates online.

The rest of the book explores the possibilities of designing technologies as *democratic mediums*. This begins with case studies in two very different attempts to design a participatory society without violence at its foundation: the transformative justice movement working toward police abolition and the "BUIDL" culture surrounding the Ethereum blockchain. From there, I call for designing toward *governable stacks* at the level of communities. Stack design can draw at once from a new kind of software paradigm, *modular politics*, and an approach to learning from the breadth of human experience, *governance archaeology*. Finally, I consider how *governable spaces* might be the basis of a fresh orientation to policymaking in various contexts.

Throughout, I present brief profiles of projects that have come out of the Media Economies Design Lab, which I lead at the University of Colorado Boulder. These are proofs of concept more than polished products. Through them I have sought to hold my ideas accountable to communities of collaborators and to code that runs. Consider them tangible gestures toward how the ideas here can come to life in practice.

It should be evident by now that this book comes with limitations. I have written it primarily with fellow researchers and other obsessives in mind, not as an introduction to online governance or a how-to manual. Other publications, including others of mine, will be more accessible for some readers. Those who conflate governance with governments will come away disappointed, as this is a book about spaces that often do not map cleanly on to territorial politics. I also hold in suspension a matter that concerns many scholars of governance: the relative efficacy of various types of governing regimes, democratic or otherwise. Any kind of governance among humans will involve contradictions, crises, and failures, and all the more so when governance takes new forms. I defer questions of efficacy—and, further, I reject their promises as deceptive—until the spaces at hand have greater capacity to define their own goals against which efficacy might be measured. Therefore my pursuit of democracy is not so much analytically utilitarian as plainly a priori—how can we settle for anything else? This book dwells largely in the negative space of neglect, of what has not been adequately tried or even imagined.

Despite presenting an argument optimistic for participatory politics, I am sympathetic to recent critics of widespread participation as burdensome, elitist, or conducive to uninformed governance.[43] A world of many governable spaces online could present an overwhelming burden to a user simply trying to access multiple services. Most users will lack a sophisticated grasp of the platforms they inhabit, if only because they use more than they have time to adequately understand. The self-governance I call for must be tailored to the context—sometimes highly participatory, other times relying more on trusteeship or representation, jury-like sortition or even market-based prediction. At the end, I will gesture toward the need for governance designs sensitive to economies of attention. Governable spaces must calibrate what they expect of people to a condition of *metagovernance*, of traversing multiple, plural governance environments in a way that is sustainable, tolerable, and comprehensible. What doing so requires, at this writing, I can only guess.

In between these shortcomings, I hope to provoke a more widespread recognition that the design of everyday self-governance in online spaces matters. But much of what I argue for remains, by necessity, untested supposition. The rehearsal stage for online self-governance has yet to be built. I hope to motivate its construction.

Implicit Feudalism

The Origins of Counter-democratic Design

There is a peculiar kind of structure that appears when online life takes institutional form.

In a statement published on November 30, 2020, ten Black Lives Matter chapters in the United States and Canada declared, "It is time for accountability."[1] The statement raised questions about what had become of the chapters' parent organization, the Black Lives Matter Global Network. It noted that Patrisse Cullors, who first posted the #BlackLivesMatter hashtag on social media in 2013, had become both the sole board member of the network and its executive director. The chapters' statement was concerned with transparency and participation surrounding the direction of their shared movement, as well as their own lack of financial support. The following year, Cullors stepped away from the organization. But a strange fact remains, a flagrant deviation from the norm of nonprofit board governance: in the waning days of 2020, a year when Black Lives Matter had become a historic anti-racist uprising across the country and the world and absorbed tens of millions of dollars in donations, its flagship organization had only a single board member.

Consider, then, a very different sort of singular leader. On February 1, 2012, Mark Zuckerberg issued a letter to investors ahead of the initial public offering for Facebook, the company he founded out of his Harvard dorm room. In it, he introduced "the Hacker Way," Facebook's "unique culture and management approach" based on being "open," "meritocratic," and willing to "move fast and break things."[2] What the letter did not explain to investors, however, was the fact that the company had instituted a dual-class stock structure, ensuring that even after the public offering, Zuckerberg would retain majority control. Contrary to the norms of Wall Street, but following some other internet companies like Google, the founder would remain in charge indefinitely.

That word *founder* evokes the scene of a foundry, the precursor to the startups' dorm room or garage, the dreary place of technological invention. In a foundry, metal becomes pliant under heat, red-hot and liquid. But when it cools, the metal turns solid. Founders solidify too. They stay in place, even against the longings for a decentralized protest movement and the will to power of institutional investors. In Groups on Zuckerberg's Facebook, power works the same way: if you start it, you keep it. In Black Lives Matter, the logic of a hashtag became the governance of an organization. The politics of the foundry holds its shape in the politics forged there.

A DARK PATTERN

This chapter considers how online platforms train users to interact with each other through certain widespread interface designs. I argue that an *implicit feudalism* informs the available options for community management on the dominant platforms for online communities. It is a pattern that grants user-administrators absolutist reign over their fiefdoms, with competition among them as the primary mechanism for quality control, typically under rules set by platform companies. These practices emerged from particular technical conditions dating to early social platforms. They have since bled into widespread social and political norms. But implicit feudalism is not a necessary condition.

I do not use *feudalism* in a historically precise sense, as there is much to distinguish online communities from the medieval European regime of land tenancy and its lord-vassal relations. Rather, I use the word metaphorically to describe concurrent communities across a network, each subject to a power structure that is apparently absolute and unalterable by those who lack specific permissions.[3] I do not for the moment mean to focus on ways in which the digital economy appears to be fostering a new feudalism of wealth inequality, though the economic dimensions will become more relevant as the argument develops. Implicit feudalism is primarily a matter of software design. It is a habit: a familiar way of doing things, along with the technical debt from past designs, around which business models have grown. I also recognize the pejorative connotations that the word generally carries. It may be true that many of the feudal practices considered here have been sensible and efficient; they may be especially appropriate at certain stages of a community's life cycle, such as early on, or at moments of transition. But I cannot completely hide my disappointment in the phenomenon or my bias for something more democratic.

By *implicit* I mean that while platforms may not explicitly proclaim or seek to practice some old-world feudal ideology—to the contrary, many claim participatory and inclusive ideals—a feudalism lurks latent in the available tools that guide and limit user behavior. Despite what the tools are supposed to do, they steer us toward something else. Implicit feudalism places unnecessary limits on the possibilities available to communities, curtailing the cultivation of online democracy. An expectation typically associated with democracy, for instance, is that those subject to an authority have the capacity to transfer the authority to someone else. Even this, in our online lives, is a rarity. The mechanisms necessary for many basic democratic processes are missing under the regime of implicit feudalism.

Democratic practices can emerge among feudal technologies. Administrators may feel rhetorical or social pressure to respect the values of community members in how they exert their otherwise absolute authority. Feudal networks can thereby exhibit forms of accountability that political scientist David Stasavage calls "early democracy," resembling the councils and assemblies of hereditary chiefdoms.[4] Communities may repurpose features like emojis and polls to carry out decision-making functions. But under implicit feudalism, inclusive governance requires clever adaptations of available feature sets, against the grain of the user interface. Consequently, empirical studies have concluded that nondemocratic practices are the most likely outcome in online communities, seemingly in keeping with the sociologist Robert Michels's 1911 prediction that human societies naturally drift toward an "iron law of oligarchy."[5] Yet upon examining what the available tools allow, the observed oligarchic outcomes begin to seem preordained. Implicit feudalism has forestalled social and political questions of how community governance might otherwise occur.

To clarify the concept, I adopt a media-archaeology approach, which looks to artifacts of the past whose traces appear in the infrastructure of the present. Specifically, this means probing the ways that technical contexts of early online communities organized—and still organize—the realm of the possible. In the past, we can also find means for raising new questions about the assumptions of the present. According to Erkki Huhtamo and Jussi Parikka, "media archaeologists . . . construct alternative histories of suppressed, neglected and forgotten media that do not point teleologically to the present media-cultural condition as their 'perfection.'"[6]

I undertake a close examination of historical documents, the machines and cultures that accompanied them, and the afterlives of both in the machine-mediated practices that surround us today. Early on, feudal governance catered to the technical circumstances of the platforms, as well as to the offline legal forms of ownership and control over their hardware. Corporations found benefits in encouraging centralized control among user communities. Those benefits then informed the designs of later technologies, later business models, and later cultural norms. Artifacts of even pre-internet experiments are still buried in the soil upon which online cities have been built, still lodged in their foundations.

What kind of concept is implicit feudalism? One way to think of it might be as a species of social-media affordance. According to a comprehensive literature review on the topic, social-media affordances are "the perceived actual or imagined properties of social media, emerging through the relation of technological, social, and contextual, that enable and constrain specific uses of the platforms."[7] Implicit feudalism enables its admins' authority while constraining what users in general can practice and even devise. Yet it falls short of affordance status to the degree that it is not something people ordinarily perceive or imagine. Social-media platforms do not advertise implicit feudalism as a feature; users do not often demand or notice it. Rather, it lurks in what the affordances lack, a negative space outside users' experience with platforms. When we do notice it, implicit feudalism appears as a disaffordance: a field of actions that platforms seem to inhibit. But in typical online life, it is merely a willingness to accept and a failure to question systems with impoverished feature sets.[8]

I hope this chapter aids in unraveling that acceptance. Noticing implicit feudalism is the first step toward making it less ubiquitous. I follow in the footsteps of the "#darkpatterns" campaign among user-experience professionals, which seeks to dissuade peers from disingenuous techniques that "trick users into doing things" against or without their will.[9] Dark patterns might sneak a monthly subscription into what users assume is a one-time donation or encourage sharing excessive personal information or make simple acts like unsubscribing unreasonably hard. Compared to these, implicit feudalism is a creature of habit more than of malice.

Let me add to the conceptual cauldron the classic distinction of economist Albert O. Hirschman between the signals of "exit" and "voice" in organizational life.[10] Exit is the capacity to depart, such as by quitting a job or shopping with a competitor; voice is the capacity to make change from within, such as by lobbying one's city council for a local policy change or filing a complaint about a defective product. If one doesn't like how an online community is being run, one can complain, too, but one's primary recourse is exit—to choose another community or create another in an open market. Yet, as in other kinds of markets, the social costs of exit can be higher than they appear from a purely technical standpoint.[11] The button to leave is always there, but actually using it might incur personal or professional costs. It might mean losing friends or access to one's culture.

Online spaces do support certain kinds of voice. They excel at chatter. Social media have facilitated a golden age of complaint against every imaginable authority, from corporations and politicians and teachers to the overworked volunteers trying to moderate posts from a thousand strangers. Seth Frey and I have therefore argued for the need to refine Hirschman's distinction with a more hair-splitting one: to distinguish *effective* from *affective* voice.[12] Affective voice can be heard in the maelstrom of online emotion and persuasion that flows so freely. It is at least the appearance of freedom; users can speak out and affirm each other into virality. But they must wait for admins or whoever else holds the keys to act on their complaints. Effective voice, meanwhile, is the voice that the peasants lack under feudalism, the instrumental power to change something, whether the nobles like it or not. We defined the effective sort of voice as "individual or collective speech that brings about a binding effect according to transparent processes." This might be the ability to vote out an admin, for instance, or to form unions among users or to require that moderators have to follow rules like everyone else. These are basic features of so much institutional life in democratic societies, at least before it all went online. There, for most of us, effective voice is mostly absent.

This chapter presents a genealogy of implicit feudalism in online communities, chronicling its emergence among particular network structures before and during the early internet. These appear to feed directly into the designs of more recent platforms for online communities, from collaboration tools to corporate social media. From there, implicit feudalism shapes our practices of governing and problem-solving, seeping outward from technical particulars to our social worlds.

ARCHAEOLOGICAL SITES

The excavations that follow reveal a sequence of software designs, together with the technological and cultural norms that accompanied them. The choices of examples are selective—genealogically significant, I argue, but inevitably incomplete. Similar patterns also occur in video live-streaming, question-and-answer platforms, multiplayer gaming, and productivity software, but I do not dwell on them here. I neglect, for now, parallel stories that occurred outside the US software industry, such as state-led social networks in Europe and the social platforms behind China's "Great Firewall." But even within these relatively narrow bounds, there is much to unearth.

Progenitors: BBS, Usenet, and Email Lists

Online bulletin board systems first appeared in the late 1970s, offering computer hobbyists outside academia and military-funded research centers their first experience of digitally mediated community.[13] The internet did not yet exist. BBSes typically resided on a single user's computer at that user's home, running one or another variant of specialized, customizable BBS software. Users' computers could log in through a phone line, post messages and files, and download content others had posted. The user who hosted a BBS became known as a "sysop," short for "system operator." Interviewees in a film called BBS: The Documentary testify to the intimacy of the sysop experience. One sysop describes lying in bed and being able to infer what users were doing on the BBS from the sounds of the computer on the other side of the room. Many sysops thus regarded users as guests in their homes, resulting in both generous and domineering behaviors. "This bulletin board is in my house," a sysop in the film imagines declaring, imitating his more prudish peers. "I will not have any swear words on it!"[14]

With hospitality came power. As a sysop, says one informant, "you could do whatever you wanted." Says another, "At the end of the day, it is the sysop who is

FIGURE 2.

the ultimate judge, jury, and executioner"; after all, the sysop could say, "If you don't like it, get off my computer, get out of my phone lines!" All rights emanated from the sysop.[15]

Media historian Kevin Driscoll recounts how sysops found themselves becoming not just hosts but lawgivers through their unique relationship to the system: "They were the makers and enforcers of social policy. Ultimately, the sysop possessed a form of total authority because they lived under the same roof as the host PC. In a moment of frustration, the sysop could always pull the plug and shut down the whole system." While a sysop's absolute power stemmed from the power to terminate the community, users had power of their own stemming from the option to exit—to leave one BBS for another: "If a user or group of users found themselves in an unresolvable conflict with a sysop, they were always free to depart and create their own system. The freedom for users to leave the system created a check on sysops' power and created a sense of mutual accountability within the community."[16]

The first specification for BBS software describes the system operator (not yet a sysop) as a technical functionary, performing maintenance on the machine and using special message-deletion privileges in cases of user carelessness.[17] But before long, the not-under-my-roof spirit infused the feature set of BBS software, granting sysops close-grained authority to sanction and censor users. There were also more democratic options available, such as the ballot-counting votemgr program for the FidoNet BBS network. The OneNet network of BBSes went so far as to have a constitution and board structure; the founder, Scott Converse, once told me the story of when the members voted him out of power.[18] But for the most part the ownership of the non-virtual hardware bled into virtual feudalism.

One important motivation for sysop absolutism was legal liability. However fun it might be to imagine virtual spaces as indifferent to the world outside, BBS guides came with frequent reminders that the owner of the machine could face consequences for what the users posted. Even *The Anarchist's Guide to the BBS* admits, after sympathizing with those who might want to talk about "bombing the local embassy of some country that you don't like," that "the bottom line is that you may well be responsible for anything that happens on, or as a result of, your board."[19] Such concerns made it a norm for sysops to verify even pseudonymous users with phone calls or mailed documents, ensuring that their control over the virtual system could extend to users' offline identities.

A case in point was LambdaMOO, an all-text online world where users interacted with each other in the rooms of a virtual house. LambdaMOO became notorious for being host to a "rape in cyberspace," a prolonged case of textual sexual assault that Julian Dibbell recounted in a *Village Voice* feature.[20] Crises of bad behavior resulted in forays into user governance, such as a petition-based system for setting and enforcing rules. But handing users power began to unnerve the administrators of this social experiment—which was housed at Xerox PARC, a corporate research entity. In 1996 the admins announced that they were "reintroducing wizardly fiat" with veto power over user self-governance, due to the realities of non-virtual jurisdiction: "So long as the MOO is located on a single RL [real-life] machine at a single RL site subject to RL laws and liabilities, there will be those deemed responsible for the use of that hardware."[21]

In realizations like this, we find a formative moment of implicit feudalism. Regardless of whatever limitless possibilities seem to exist in virtual space, if that space lives on someone's server, then the possibilities end at what that someone, together with the legal regime where they live, will tolerate.

In 1980, another approach to networked community appeared in the form of Usenet.[22] Like a BBS, it was a forum for asynchronous content-posting. But rather than residing in a sysop's home, Usenet distributed its "newsgroups" among interoperable servers, typically hosted by universities or corporations. A vibrant, even anarchic culture emerged among users, as they reveled in the opportunity to create communities far beyond the telephone area codes that typically circumscribed BBSes. But as the host organizations took stock of the free-for-all inhabiting their computers, they sought to establish discipline.

The structure of Usenet's network had developed into a hierarchy, with most Usenet providers relying on a small number of central servers to circulate content. The central sysadmins became known as the "Backbone Cabal," and they instituted the disciplinary reform. In the "Great Renaming" of 1986, Usenet's major public spaces came under the authority of an organization eventually known as the Big 8. The Big 8 still governs key sections of Usenet. There is a voting system for adding new newsgroups, though this political process is not binding over technical power; in some instances, sysadmins have simply refused to carry newsgroups approved by a vote.[23]

The board of the Big 8 is self-perpetuating, meaning that current members choose future members. Similarly, moderators of particular newsgroups choose their own successors according to processes specified in a Big 8–approved group charter or, if there is a break in the line of succession, by the Big 8 board. Once chosen, a moderator's power is much like that of a BBS sysop. According to one Big 8 documentation page:

> *Who can force the moderators to change their policies?*
> · Nobody.
> *Who can force the moderators to obey the group charter?*
> · Nobody.[24]

The document continues:

> *Why won't you give us more help with our group?*
> · The group belongs to the moderators and the users.
> · Usenet is not structured in such a way that outsiders can intervene.

Usenet's governance was robust enough to foster a popular set of online communities that, more than the homebound BBSes, served as a virtual public square both before and after the rise of the internet.[25] The circumstances of operating on a shared network and taking up shared server space required governance mechanisms capable of at least some collective decision-making. But at the level of most user experience, feudalism reigned. One study of Usenet's evolution used that language explicitly: "The system's initial democracy and egalitarianism had been replaced by a feudal structure, in which system administrators deliberately, if self-mockingly, referred to themselves as 'barons' (and to users as 'serfs')."[26]

Among people with access to the ARPANET and the early internet, email was the "killer app"—the use case that made the technology truly useful. As email became a medium for communities, it became another site of implicit feudalism.

By the mid-1980s, the email discussion software ListServ began replacing Usenet on university systems.[27] It enabled institutional sysops to fully control the media of conversation. Other email-list programs, such as Sympa and Mailman, emerged later in the 1990s. Google Groups, both an email-list platform and a gateway to Usenet, appeared in 2001.

Email lists can take many forms, ranging from announcement lists to moderated or unmoderated discussion lists. But implicit feudalism governs every major email-list system. Lists have particular admins or moderators, beginning with the list founders and followed by whomever they appoint. Organizations such as universities can oversee the lists they allow on their servers, much as the Big 8 board does for Usenet. But beyond that, email-list software grants list admins full authority over such matters as list membership, posting rights, and documentation about the list's purpose and policies.

The feudal power structure inculcates cultural norms. According to a widely circulated post on a computer-security email list, "Mailing lists should be run as an autocracy with the admins/owners as the rulers and the charter as the law." The author regards the notion of "self-moderation" among users as an "ill-considered and badly implemented mockery of a democratic process [or witch hunt, depending on your perspective]."[28] By this account, sure: the autocracy in a well-developed list may have a constitution-like charter that lays out certain rights and responsibilities. But any such feature is extraneous to the software and the power it assigns. If a charter places an obligation on the moderators, only those same moderators can enforce it. This is by design, in order to protect the administrators of the servers on which the software runs and their bosses who own those servers. Feudalism, once again, is a practical outgrowth of underlying conditions.

Concurrent with the development of asynchronous discussion spaces were more synchronous community tools, which fall under the general rubric of "chat." These date, for instance, to features in the educational PLATO system that gained traction in the early 1970s.[29] One of the most important, persistent examples is the Internet Relay Chat protocol, or IRC, first developed in 1988; in addition to its widespread use among technologists and hobbyists of various stripes, IRC prefigured many aspects of more recent, centralized, and commercial chat platforms like Slack, down to its channels marked with a hash symbol.

The organizational structure of IRC resembles that of Usenet, with networks of independently operated servers providing access to a shared set of resources.[30] IRC gave rise to a system of network operators and channel operators, or chanops, the latter of which have moderation privileges over particular chat rooms akin to those of sysops and email-list admins—setting basic rules and enforcing them by removing users. Although in principle anyone can create a new network or a channel on a public network, in practice most IRC activity occurs among a small number of the largest networks. Channels with iconic names tend to become canonical, making exit rarely feasible; quilters will always drift to #quilters, regardless of its moderators' track records. Within channels, IRC permits the additional possibility of bots, or software-defined users that assist operators in tasks useful for an always-on, synchronous system—ranging from issuing reminders about a channel's topic and rules, enforcing those rules, or even merely staying in a channel to prevent another user from claiming control over it. These bots presaged aspects of more sophisticated algorithmic governance in the implicit feudalism to come.

Beneath all the early networks were the flows of power in computer systems themselves. The networks reiterated the structure of their technical substrates. The design of UNIX-style operating systems, for example, prioritizes the pursuit of modular neutrality, which media scholar Tara McPherson likens to the social systems that perpetuate racism through the intentional blindness of compartmentalization.[31] All user permissions derive from those granted through the "root" user—a professional administrator if it is a corporate system or else simply the computer's owner. Tools such as the popular database software MySQL use the language of "master" and "slave" to describe relationships in the software.[32] When machines turned into servers on a network, the root-master monarchy became the networks' politics. As people began to make communities on these networks, creating and producing on them, the computer's way of granting power through permissions became the default social order.

Contributors: Commons-Based Software and Wikipedia

An often-celebrated source of democratic promise in internet culture is what Yochai Benkler dubbed "commons-based peer production":[33] users coming together online as peers to collaborate on projects. Especially remarkable is how the Free Software and Open Source movements—which I will refer to collectively as *open source*—produce billions of dollars' worth of software each year that anyone can freely access and modify. The success of projects such as the Linux kernel and Wikipedia indicate that peer production is capable of producing scalable, reliable infrastructure. Yet democracy is only occasionally part of the process.

The Linux governance model centers around founder Linus Torvalds, who wrote the first version of the software while still a student in 1991. He is popularly referred to as the project's "benevolent dictator for life," or BDFL.[34] In theory, anyone can contribute code to Linux, but Torvalds holds ultimate power over what ends up in the releases. Notwithstanding a 2018 sabbatical "to learn how to stop being an asshole," as one journalist put it,[35] he has remained in power all along. His role in perhaps the most influential open-source software project is indicative of how implicit feudalism has helped produce a culture of explicit dictatorship. This occurred more as a result of omission than ideology. Most open-source communities have avoided explicit governance, regarding it as a distraction from writing code. The result was a cascade of power vacuums, which implicit feudalism stood ready to fill.

Git is the version-control software that Torvalds first built in 2005 to manage the development of Linux. It enables developers to track revisions in a project and integrate the changes from many contributors. It has since become the ubiquitous collaboration tool for open-source projects. On its own, Git seems to break the norm of implicit feudalism. No one developer's version is intrinsically canonical, so every user becomes in some sense an admin, a first-class citizen. But this means that Git leaves a power vacuum. Developers must eventually choose a canonical version of the code to be the basis of any official release. Somehow they need to fill the vacuum and decide which version to publish. Torvalds filled the vacuum for Linux with his BDFL status—quite simply, he decides which version is canonical and which community contributions it includes. Linux and many other projects employ email lists for the discussion and decision-making. The implicit feudalism of the list supplies the politics that Git lacks. Whoever controls the list controls the software.

Today, Git is most widely used through hosted platforms, particularly GitHub, a commercial service that Microsoft purchased in 2018 for $7.5 billion. GitHub embeds Git into a social network that fills the Git power vacuum. A familiar access and permissions system identifies "owner" and "collaborator" roles for any project. The creator of a new project begins as its owner and remains so until assigning someone else to that role and relinquishing it. External users can also "fork" a copy of the project, edit it, and either submit their changes back to the original or attempt to release a competing version. Exit is therefore at least in theory possible, and users can make their voices heard by posting in discussion threads called "Issues." But the effective voice lies with the owner and the owner's delegates. Unlike Git on its own, GitHub establishes a canonical version of the code for any given project, managed by its permissions system. GitHub fuses Git with a feudal governance model.

Widespread abuses of power in open source—for instance, Linus Torvalds's notoriously rude treatment of developers—helped give rise to codes of conduct for software projects that seek to limit the scope of acceptable behavior and specify the responsibilities of admins.[36] At first, leaders of prominent software communities rejected the idea that more explicit rules were necessary, but the persistence of developers like Coraline Ada Ehmke forced projects to recognize that power vacuums were not acceptable, particularly as cases of sexual harassment mounted.[37] Linux itself has adopted Ehmke's now-popular code of conduct, the Contributor Covenant, and GitHub encourages project owners to adopt a code of conduct as well. Using a code of conduct on the platform, however, depends on the project owner's willingness to adopt, abide by, and enforce it.

Less feudal approaches are evidently possible. The Debian Project, which produces an important Linux-based operating system, self-governs as a kind of liberal democracy.[38] Its Debian Constitution specifies procedures including the election of a "project leader" by Debian's developers. Skilled developers join the organization through a detailed and meritocratic on-boarding process. But in its formal republicanism, Debian has been mainly an outlier. Much commons-based software development occurs under the power of a particular benevolent dictator or a hierarchical company.[39] Democratic arrangements appear only occasionally, usually among more developed software communities such as Debian and the Apache Software Foundation, whose developer-members elect their nonprofit organization's board. Apache has a rule for its hosted projects: "No dictators or corporate overlords are allowed."[40] Perhaps it helps that both Debian and Apache operate on a nonprofit basis, rather than being beholden to corporate imperatives, although Linux operates through a nonprofit foundation, too.

In perhaps the most famous example of online peer production, the nonprofit encyclopedia Wikipedia operates through a sophisticated system of self-governance among active volunteers. Wikipedia also possesses a benevolent dictator in the person of founder Jimmy Wales, who prefers the metaphor of "constitutional monarch."[41] Wales oversees a complex of tiered roles, open participation, and elec-tioneering from a "founder's seat" on the Wikimedia Foundation board, although his powers have diminished over time after several cases of overreach.[42] Any user, in principle, can ascend the ranks of influence and position—holding such roles as "administrator," "steward," and "bureaucrat." Users are elected to these roles by their peers. The outlier on English Wikipedia is the role known as "Jimmy Wales." According to the website's documentation, "Jimmy Wales holds a special role in the governance of the English Wikipedia, due to the central and vital stake he had in its founding. This authority is used on an ad hoc basis, when other decision-making structures are inadequate or have failed in a particular situation."[43]

One of the "Five Pillars" of Wikipedia is that "Wikipedia has no firm rules," but contributors have assembled a formidable assortment of policies on dozens of subjects. Aside from some external email and chat forums, most of the platform's governance occurs on the editable pages of Wikipedia itself, formatted according to certain norms. It is a remarkable instance of "eating your own dog food"—an organization using its product in the process of making that same product. Wiki-pedia's governance also exemplifies how much extra work it can take to depart from the dominant pattern of implicit feudalism.

The open-source software underlying Wikipedia, MediaWiki, is in principle available for others seeking to replicate the famous encyclopedia's success. However, without long-cultivated norms around the use of "Talk" pages and a complex system of permissions and roles, the software itself offers little in the way of democratic tooling. Like most Web-based platforms, a new deployment of MediaWiki grants privileges solely and completely to its administrator. It is not therefore surprising that in a study of 683 MediaWiki-based deployments on the commercial platform Wikia, most use cases tend toward oligarchic governance.[44] Without Wikipedia's deliberate cultivation of democratic and bureaucratic process, the software facilitates a long tail of feudalism. Governance on Wikipedia itself has drifted toward less inclusivity and dynamism over time.[45]

In principle, the power vacuums that software designs leave open could allow for diversity and healthy self-governance. But as feminist activist and scholar Jo Freeman famously observed, a "tyranny of structurelessness" frequently arises—one in which the absence of an explicit hierarchy in a system results in an hidden, difficult-to-alter hierarchy imported from external social forces.[46] Freeman's essay, first written for feminist "rap groups" of the early 1970s, has found an afterlife among those in tech culture who recognize tyrannies of structurelessness around them. As Zeynep Tufekci puts it, "The tyranny of structurelessness has merged with the tyranny of platforms."[47] If groups do not develop intentional "democratic structuring," Freeman argued, informal power structures will form, usually reinforcing existing hierarchies and privilege. The notion that "anyone" can contribute to and even co-govern an open-source project—a notion sometimes referred to as "do-ocracy"[48]—fails to recognize that not everyone is equally equipped with the free time, knowledge, and incentives to participate. Power vacuums can produce the most entrenched feudalism of all. Among the "base assumptions" of the San Francisco feminist hackerspace Double Union is that "meritocracy is a joke."[49]

The Rise of Platforms

As Facebook's public relations apparatus was beginning to come to terms with the platform's contested role in the 2016 US election, Mark Zuckerberg issued a lengthy essay called "Building Global Community." In it, he indicated a turn toward emphasizing "meaningful groups" over the user-curated political news that was making Facebook notorious. Recognizing the limits of the company's regulatory capacity, he mused about the opportunity to "explore examples of how community governance might work at scale." The essay contains various nods to US political pieties, including a quotation from Abraham Lincoln; at the time, some observers speculated that Zuckerberg might be considering a run for the presidency.[50]

At least from a technical perspective, the rise of globe-spanning corporate networks presented an opportunity for departing from implicit feudalism. No longer was a community's virtual space sitting in somebody's house or on a university server; now, the infrastructure was in the hands of companies that described their product as "platforms." The term bears a claim to neutrality, to simply providing an empty stage for users to fill.[51] Seemingly, the platforms created a new layer of abstraction: compared to earlier systems, communities form at a greater remove from the servers. In 1996, the US Congress passed the Communications Decency Act, whose Section 230 protected platforms from most liability for user behavior.[52] The companies could control the platform layer, while enabling communities to govern however they liked. Yet implicit feudalism persisted, even as platform founders preached democracy.

Facebook is the world's largest private social-media network, with around 3 billion active users. It has enabled communities to form with its Groups feature since 2005, the year after the website first appeared. Reddit also began in 2005, and by 2008 the social-news platform came to be organized around user-created and user-governed groups known as "subreddits." Reddit's active-user population is an order of magnitude smaller than that of Facebook, which still places it among the top ten US networks. In many respects, the two platforms are quite different; Facebook emphasizes users' "real names" and mutual connections, while Reddit tends to rely on individualized, pseudonymous identities marked with reputation-based "karma." Both enable significant degrees of local control among user communities, in distinct ways. They have become spaces of tremendous creativity and democratic practice. Nevertheless, both adopt and further advance the pattern of implicit feudalism inherited from earlier networks like BBSes and email lists, despite lacking many of their predecessors' technological constraints.

Why do feudal defaults persist on large platforms? A Facebook Group doesn't reside in its creator's house. A subreddit doesn't consume the computing resources of its moderators, only that of Reddit itself. It is no longer so obvious that the founder of a community should have dictatorial say over it. The norms and design elements of implicit feudalism are no longer a matter of technical necessity. But they became a business model.

Managing online communities can be hard, thankless work, involving negotiations with an often tiny minority of disruptive users and reviewing potentially traumatic content so that others don't have to.[53] One of the first large commercial platforms, America Online, began appointing "community leaders" in the early 1990s to moderate its chat rooms and message boards in exchange for reduced cost of access, providing compensation for what was generally perceived as volunteering. But some of these people recognized that their efforts were generating real profits for the company and began to protest; the program drew scrutiny from the Department of Labor as under-compensated work.[54] Since then, platforms have avoided such gray-area compensation. Instead, the allure of implicit feudalism has served as another kind of compensation to incentivize the labor of community management. Rather than criminally low wages, platforms offer moderators the perk of unchecked power.[55]

An exception that proves the rule among social platforms is Slashdot, an early social-news website with a tech-savvy user-base. As Slashdot grew during the late 1990s, it developed a complex system of moderation (and "metamoderation") based on a "karma" score—the term Reddit would later adopt.[56] As users accrued karma from other users, they gained the power to moderate and evaluate others' moderation decisions, producing a basically functional, Wikipedia-like culture of responsible voluntarism. Reputation became a kind of compensation. Slashdot thus employed a fluid system of mutual endorsement rather than a Debian-style electoral republic, but it similarly showed that an open, dynamic system of user empowerment could manage the content on a large platform in ways that generally satisfied its users. Perhaps such a model was even too responsible, failing to produce the kind of provocation and engagement that commercial social networks thrive on.

One mechanism of apparent self-governance that appears in both Facebook and Reddit is the ability for non-moderator users to evaluate fellow users' posts—on Facebook with the Like button and its various affective sub-options and on Reddit with "upvotes" and "downvotes." These tools allow users to mutually decide which content is more worth each other's attention and thus which should rise to the top of the group's feed. The platforms also allow users to add comments, which have amplifying effects as well. But the most definitive powers of amplification (elevating messages to the top of a group's feed) and sanction (ejecting posts and users) are reserved for those with administrative roles, who gain their authority by appointment and succession deriving from the group's founder. Interviews with admins on both platforms reveal that they rarely consult with non-admins on decisions about how to use these powers. Ordinary users' evaluative tools thus seem to operate as assists on behalf of admins, as well as the companies' business interests, more than as a means of shared governance.[57] The strongest form of effective voice for ordinary users remains that of exit: to leave a given Facebook Group or subreddit for another or to start a new one.

Facebook and Reddit implement advances in implicit feudalism over earlier paradigms. For instance, rather than merely offering blank text fields for rule-making, as in MediaWiki and GitHub, these platforms have developed structured rule-making interfaces for group admins. Artificial intelligence tools, such as Facebook's "false news" detector and Reddit's programmable AutoModerator,[58] offer to streamline the labor of moderating content. Analytics dashboards present admins with detailed reports on the activity of their groups, in effect gamifying the admin role toward maximizing user usage. Such tools add to the panopticism and potency of implicit feudalism's repertoire.

Feudal community governance has become a norm in the governance of platform companies themselves. This is most evident in the power Mark Zuckerberg retains over Facebook through its dual-class stock structure. To extend the metaphor of feudalism: if admins are ladies and lords, Zuckerberg acts as a monarch, who holds similarly absolutist powers over the rules by which his nobles operate, even without appearing to interfere in their fiefdoms directly. Zuckerberg also rebuffs shareholder proposals to put constraints on his authority. Yet Facebook has meanwhile engaged in "democracy theatre," such as its 2009 user referendum on proposed changes to its terms of service.[59] For users' votes to be binding, the company stipulated that 30 percent of its over 1 billion users at the time would need to participate—a scale equivalent to the entire US population. As one might expect for an unprecedented process on a decision about complex legal language, well under a single percentage point of the quorum was reached.

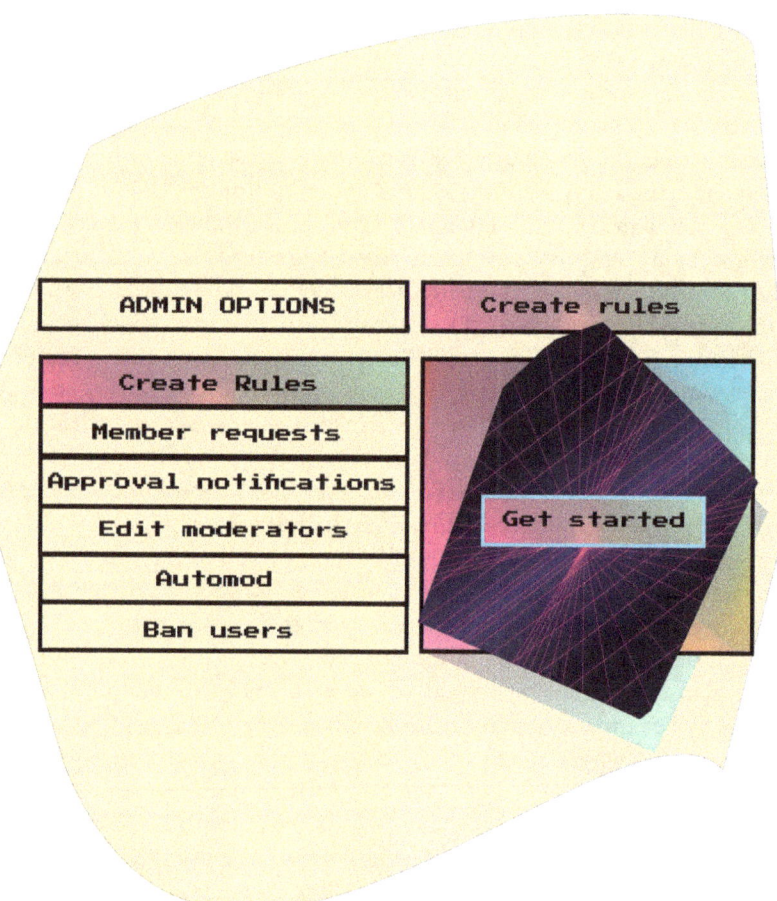

FIGURE 3.

The company shrugged, called the vote "advisory," and proceeded with the rule change as it saw fit.

Reddit's corporate edifice has had its own brushes with a kind of democracy, such as in the 2015 "Reddit revolt," when moderators galvanized by crackdowns on toxic behavior during the Gamergate controversy turned on the company. They switched their subreddits to private en masse, resulting in a widespread blackout of the platform's content and the resignation of interim CEO Ellen Pao. With the victory, however, came heightened enforcement of site-wide policies that brought about more conformity between the platform's policies and moderator policies at the subreddit level.[60] The moderators can lord over their fiefdoms, but they face consequences if they try to band together against the monarchy.

Conway's Law is a celebrated truism in software development: technical systems tend to resemble the communication structures of the organizations that create them.[61] Among companies like Facebook and Reddit, the influence has seemed to go the other way. The communication structures of technical systems informed what seemed plausible and practical for the architecture of corporations. Implicit feudalism made its way from the server permissions and the online community to the boardroom.

The centrality of implicit feudalism to online experience has at times wavered, only to return again. In a follow-up missive to "Building Global Community," Zuckerberg pivoted from a vision of Facebook as a community-oriented "meaningful" space to that of a "privacy-focused" platform for private chat and "intimate" group exchanges.[62] It was a retreat from his aspirations two years earlier for "global community." Already, Facebook-acquired platforms WhatsApp and Instagram were making gains against the company's namesake product. The photo-sharing app Instagram did not initially enable persistent groups; WhatsApp permits them within the logic of chat, as opposed to Facebook's forum-like threaded discussions. Zuckerberg appeared to be learning from China-based WeChat and TikTok in enshrining networked individuals rather than a network of communities as the rubric for platform society. TikTok in particular has shown the possibility of targeted advertising based on personal viewing habits alone, without need for a social graph.[63] This shift trades feudalism—which presumes community, however hierarchical—for platform-mediated experiences, apparently detached from any particular kind of politics. But politics seemed likely to return with Zuckerberg's next pivot in renaming the company as Meta, proposing to provide the infrastructure for entire immersive worlds. Meanwhile, ascendant community platforms such as Slack and Discord explicitly imitate the social software that gave rise to implicit feudalism—down to the "#" marking channel names following IRC and Discord's "server" nomenclature for its virtual groups. As corporate teams and mutual-aid activists alike adopt these tools as the basis of their organizing, feudal designs continue to grow in influence.

FEUDAL DEFAULTS AND THE POSSIBILITY OF VOICE

Implicit feudalism has reigned over the dominant platforms for online communities so far, from the early BBSes to Discord. Peer-production practices surrounding open-source software and crowdsourcing also exhibit it. In summary, implicit feudalism's recurrent characteristics include

- control over communities residing in an individual or a small group,
- authority deriving from founders and their appointed successors,
- opacity of policymaking and decision-making processes,
- suppression of user voice as a basic privilege of authority,
- user exit as the most forceful means of dissent, and
- sole recourse to platform owners in disputes.

While these made a specific kind of sense in the context of a BBS running in a sysop's home, that is not necessarily the case in the context of a global, multibillion-user platform like Facebook. One can just as easily imagine implicit democracy operating there as implicit feudalism. Yet the feudal pattern has by and large been written into the default behaviors of online-community platforms. Feudal powers became part of the business model, incentivizing the unpaid labor of moderation and community building. Some communities, like Debian and Slashdot, have bucked the trend and painstakingly crafted more democratic processes. But most seem to have simply gotten used to feudalism, developing their cultures and expectations around it.

Under this regime, the possibilities for community governance are constrained. Opportunities for affective voice enable users to feel heard enough to impart a fleeting satisfaction, but those opportunities rarely include the force of effective power. As anthropologist Christopher Kelty puts it, in twenty-first century digital cultures, "participation is more often a formatted procedure by which autonomous individuals attempt to reach calculated consensus, or one in which they experience an attenuated, temporary feeling of personal contribution that ends almost as soon as it begins."[64]

Governance defaults in offline domains present an instructive contrast. Even quite autocratic governments at least carry out performances of democratic institutions, such as elections and judicial oversight, because those practices have come to stand as prerequisites for legitimate authority. Regulators expect public corporations and nonprofit organizations to have governing boards that represent specific stakeholders—generally shareholders and donors—together with certain transparency requirements. Civil-society organizations such as industry associations and fraternal societies often practice at least a semblance of choosing leaders by a ballot among members. Although these mechanisms of offline participatory governance can mask oligarchy or autocracy in practice, their ubiquity makes it striking that no major online community software platform offers purpose-built features to support them. Adopting conventional democratic mechanisms online requires working intentionally and persistently against the grain of implicit feudalism.

The fact that implicit feudalism is so ubiquitous does not necessarily justify its ubiquity. It is not uniquely effective for building communities. The number of subreddits with only a handful of subscribers far exceeds those that have attracted large followings. A study of user-run servers for the game Minecraft—a technical arrangement that resembles a BBS of old—found that the median lifetime of a server is eight weeks, and more than half of admins never recruit any committed community members.[65] High failure rates may not be a bad thing; online communities are relatively low-risk environments, with minimal startup costs and minimal consequences of demise. But users do not have a high opinion of moderation on the dominant platforms.[66] There would likely be benefits to greater institutional diversity,[67] including community-centered, democratic mechanisms. Albert O. Hirschman predicted that while exit-based organizational designs excel in producing variety, choice, and innovation, voice-based designs confer greater commitment and stability. A study of the GameCenter online community, for instance, observed that when the platform's "benevolent dictator" became less active over time, subgroups developed unexpected resilience.[68]

Debian and Wikipedia do not exist in isolation. They act in concert with different kinds of regimes, playing distinct and complementary roles. Hirschman's exit-voice framework predicts that different institutional logics will serve different purposes, often in concert; they are not substitutes for each other. Simply replacing feudal governance with democratic governance anywhere and everywhere could create as many problems as it solves. Instead, the exceptional cases considered here reflect conditions of institutional diversity. That diversity takes several forms.

One form is onion-like. Debian, for instance, is not a standalone operating system; it holds a particular location in a layered ecosystem. The Linux kernel, with its rigid dictatorship, lies at the center. Debian holds a critical middle space, with its democracy enabling a slow but inclusive development process that supports even older machines with limited commercial value. Above Debian sits Ubuntu, a popular operating system supported by a for-profit company, Canonical, whose founder and CEO Mark Shuttleworth uses the online handle sabdfl, or "self-appointed benevolent dictator for life." Ubuntu benefits from the inclusiveness of Debian but funnels it into a more streamlined operating system with a faster release cycle.

Participatory self-governance appears to flourish at certain niches in the software supply chain, but it may not be as well suited for others. It appears to be more likely to emerge under organizations like nonprofits or user-owned cooperatives. Already we see, from BBSes to Facebook, and in nonprofit-owned projects like Debian and Wikipedia, that community governance tends to mirror the underlying platforms' ownership structures, along with their technical infrastructures.

A second form of diversity is the combination of different power structures into one—the idea of "mixed constitution," argued for in antiquity and adopted as a "separation of powers" by the authors of the US Constitution.[69] Both Debian and Wikipedia combine electoral processes with meritocratic barriers in order to ensure that leaders are not just popular but exhibit a high level of expertise. Usenet combines some aspects of shared governance in its board with considerable autonomy among the newsgroups. Integrating multiple governance mechanisms not only helps prevent any one entity from becoming too powerful, but it enables participants with heterogeneous skill sets to make their voices heard. Some users might bring technical skills, while others bring social skills, and they may each need their own pathways for finding effective voice in their shared community.

Thirdly, governance diversity can unfold over time. The community that produces the Python programming language had a benevolent dictator, Guido van Rossum, for almost thirty years. When van Rossum abruptly resigned from the role in 2018, Python developers undertook a process to find a new governance model.[70] They proposed a staggering set of possibilities, ranging from a new dictatorship to utter structurelessness, along with various boutique systems for tabulating votes. For a community of designers, it was a feast. This process was possible, in no small part, because the developers had a social infrastructure that mitigated the inertia of implicit feudalism: the Python Enhancement Proposal system, set of processes and tools designed for proposing and adopting changes to the programming language. As a system the community perceived as both familiar and legitimate, it helped fill the power vacuum and ushered the community from a radically divergent range of possibilities to a rather sensible, conventional result: the adoption of an elected, five-person "steering council." Without the benefit of existing decision-making practices, the habit of monarchy might have persisted, or something even less sensible might have replaced it.

For Python the end of feudalism took decades, plus a sudden disruption. That need not be the fate of others. Feudal patterns have their usefulness in certain times and places, but that does not mean they should be as ubiquitous as they have become.

"THESE TOOLS ARE OFTEN BLUNT AND SENSELESS"

The more I have learned to notice implicit feudalism, the more I see its effects. A person I have known online and off, someone I long considered a mentor, came under criticism for speech and behavior that many of us in his community objected to. Call him Miguel, though that is not his name. As Miguel experienced pushback, hostility, and lost work opportunities, he named the problem as "cancel culture." This is a label of reaction, an anxiety voiced most often among cultural elites about the threat of being "canceled"—co-opting language that began as vernacular for mass shunning on Black Twitter.[71] Miguel shared heartfelt and frustrated tales of injustice. I agreed with his critics, for the most part, and at times did so publicly. But when a letter circulated calling for "disassociation," I couldn't sign it.

What does disassociation mean? It is no clearer than canceling. Canceling is at least a playful reference to ill-fated TV shows, not a letter that you are asked to sign. How long would disassociation last, and how completely must it be performed? Could I still ask Miguel about his family from time to time, or cite the work of his that still informs mine? What if he somehow repented? There was no specified pathway to reconciliation or repair. This, I realized, was in keeping with how online life had taught us to self-organize: to rally in excess, to engage to the max, with none of the precision or specificity that stakeholders with actual power expect of each other. This is affective voice without effective voice. There was no dispute-resolution system to turn to, no way of challenging Miguel's admin status across multiple social-media spaces where his community gathered. As he protested about cancellation, he habitually removed his critics from those spaces, while they had no such recourse themselves. Some simply left on their own, but that meant leaving a community that had mattered to them. Ironically, these were communities devoted to the practice of co-governing shared resources. But the practical politics that the commercial platforms inscribed in them was feudalism.

In her book *We Will Not Cancel Us*, adrienne maree brown observes the temptation and futility of the call-out practices that have become so endemic online: "Right now calling someone out online seems like first/only option for a lot of people in the face of any kind of dissonance." She goes on: "The tools of swift and predatory justice feel good to use, familiar, groove in the hand easily from repeated use and training, briefly satisfying. But these tools are often blunt and senseless."[72]

Anxieties about cancel culture should instead be anxieties about the fact that there is no better recourse, that people feel powerless to address conflict in a pro-portionate, deliberate way. We couldn't just vote Miguel out and thank him for his service, or submit a complaint to a mediation process. To the extent that cancel culture has become a term of derision, perhaps the blame should fall not on the crowds for their excesses but on the systems that leave them little choice.

I have argued that implicit feudalism has become a nearly ubiquitous pattern embedded into the software for online communities, to the point that even appar-ent exceptions prove the rule. Implicitly feudal designs incline communities, like dark patterns, toward the iron law of oligarchy. These designs have specific, sen-sible historical origins but unnecessary persistence. Meanwhile, implicit feudalism has initiated users into a willingness to accept the exit logic and affective voice of their online fiefdoms without the effective voice of democratic participation. Perhaps the drift toward oligarchy would not be such an iron law without feudal software nudging us that way.

Recognizing implicit feudalism can have explanatory virtues. Whitney Phillips, for instance, came to recognize her study of online trolling as "a critique of domi-nant institutions" as much as of "the trolls who operate within them";[73] as with cancel culture, bad behavior may become worse in the absence of infrastructures for accountability. What other aspects of online life arise from that absence?

Implicit feudalism doesn't just lurk in the software; it reflects, expresses, and promulgates certain kinds of political habits. These habits began in spaces that appeared merely virtual, ancillary to the real politics happening in real life. But the politics of everyday online life has spread beyond the screens because it was never really contained there in the first place.

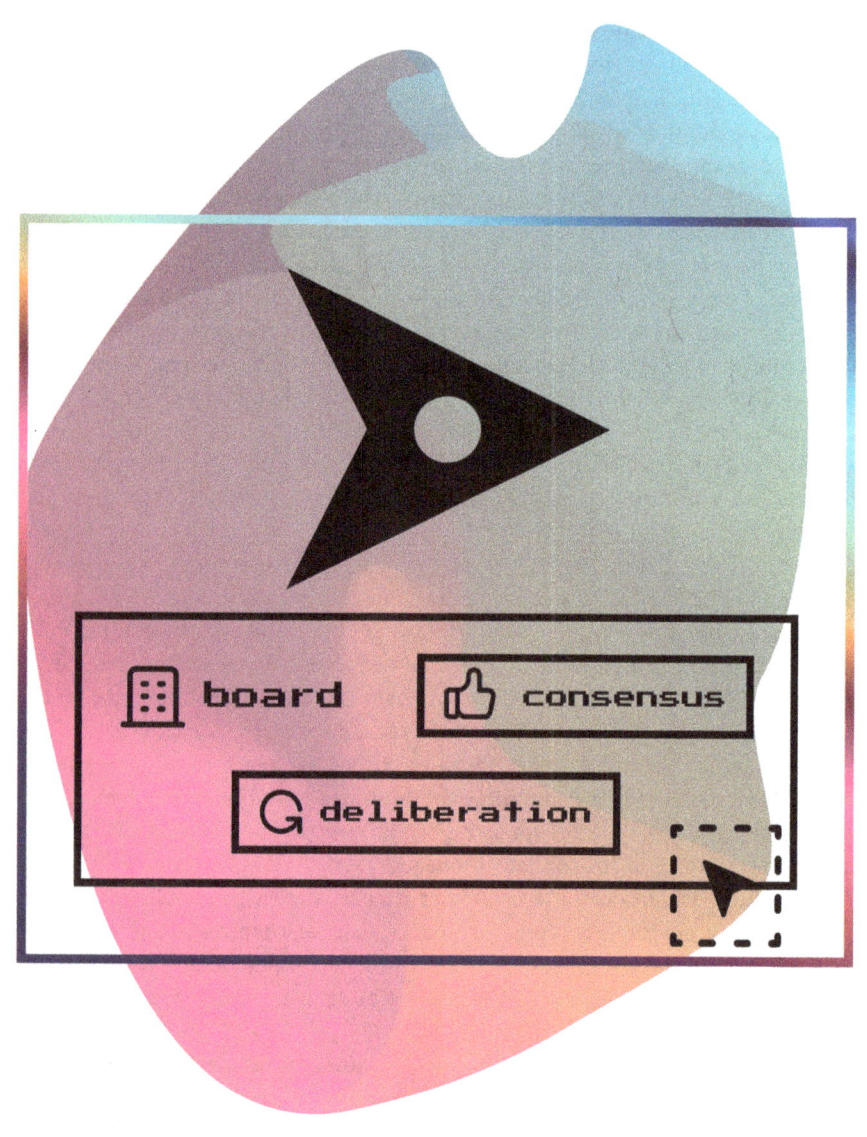

FIGURE 4.

CommunityRule

communityrule.info

Having shared rules in a group is important. They help members know when a decision is really a decision. When conflicts arise and people's relationships alone aren't enough to handle them, it helps to have a clear process for what to do. But traditional bylaws are too formal—and even too expensive to produce if lawyers get involved—for most online communities.

CommunityRule is a Web app that allows users to design their communities' rules interactively. We developed it, first, through a series of consultations with mutual aid groups that formed during the early months of the COVID-19 pandemic, as well as with open-source software communities. These helped inform our designs, and we tried to help the groups think through their processes as well.

The first version of CommunityRule was simply a series of questions that users could answer, in writing, about how their community should work. The current version enables dragging and dropping governance modules and nesting them inside each other. Modules can be configured and customized.

CommunityRule enables users to publish their rules to a public library where others can learn from them. Rules in the library can be forked—copied and modified as new rules. We have also developed a set of simple templates, reflecting several basic organizational designs, that rule authors can use as a starting point. After noticing that users found these templates useful, in 2021 we published a set of them in a free print and online booklet.

Implicit feudalism thrives on an absence of rules; admin power fills the void. More democratic communities need ways to describe the rules they want to use. CommunityRule is an attempt to imagine interfaces that make governance arrangements easy to design and understand.

2

Homesteading on a Superhighway

How the Politics of No-Politics Aided an Authoritarian Revival

Perhaps the feudal power structure of platforms for online communities could have stayed there, contained and cordoned off in virtual space. Democratic politics has long coexisted with nondemocratic workplaces and patriarchal families. The democracy of ancient Athens coincided with slavery. But virtual habits spread to other quarters of the social order. Online spaces became training grounds for other spaces. The politics of virtual life have poured over into the politics of almost everywhere else.

In the mid-1990s, Richard Barbrook and Andy Cameron published their warning about "the Californian ideology" poised to dominate the early internet: a faith that greater volumes of information and connection, fueled by capitalism, would produce a flourishing democracy. Technology could end the old partisanship of right and left through entrepreneurs "believing in both visions at the same time."[1] Silicon Valley CEOs continue to proclaim this gospel today, even as the parades of platform scandals make them do so a little more quietly. They preach that artificial intelligence will resolve the conflict of labor and capital by automating jobs. Cryptocurrency enthusiasts herald a new order in which markets can replace monetary policy. Yet the humans glaring intensely into Californian-designed devices have somehow become more polarized than we have been in recent memory. Resurgent autocracies ride Californian software into prominence and power, while democratic norms veer into precipitous decline.

The agenda of this chapter is to revisit the politics of no-politics that Barbrook and Cameron diagnosed—the culture that, according to Fred Turner, "turned away from political action and toward technology."[2] The original formulation of the Californian ideology outlined a certain kind of political economy, a social and economic liberalism capable of assailing industrial policy while tacitly relying on it. Here I turn from political economy to the micropolitics of everyday online life: how implicit feudalism encoded certain imaginations of social order into software designs, which users far from California have decoded into a neo-feudal politics.[3] I argue that the Californian ideology inscribed the habits of homesteading—a legacy so familiar, nostalgic, and violent in the American West—into the practice of online communities. Everyday experience with Californian technologies has thereby contributed to hollowing out the rudiments of democratic culture, especially the skills and habits of accountable association. These systems have aided in generating new breeds of world-historical authoritarianism. To change course, therefore, instruments such as legislation and foreign policy may be inadequate; securing a more democratic future also requires fresh attention to how online spaces organize, constrain, and enable everyday politics.[4]

My argument emerges from divergent voices and fragmentary scenes. I build on earlier critical chronicles of Californian times and places, such as those of Adam Curtis, Joy Lisi Rankin, and Fred Turner, along with intrusions from worlds away.[5] This is a story of deep mediatization, in which media become inseparable from the practice of social life and the production of culture. Throughout, I pay particular notice to cases of emergent religiosity, following Kathryn Lofton's attention to "how religion manifests in efforts to mass-produce relations of value."[6] This is because the voices I turn to repeatedly articulate or elicit diverse religious sensibilities—not a uniform religion of any sort but a cluster of interrelated appeals to transcendent forces. These appeals appear to function as mediations between macro and micro scales of social life.

Even as I begin with the Californian ideology at the center of this discussion, I decenter it. Silicon Valley, or some hegemonic subset of it,[7] has encoded its values in technologies now used the world over, but adopters have decoded meanings very much their own, which become new encodings in turn. Part of what the ideology has excelled at is disowning its history and progeny alike, an amputation I hope to deny it.

HOMESTEAD AND HOMEPLACE

Founded in 1985, The WELL became a text-only gathering place for a mixture of intellectual seekers, technology enthusiasts, and Grateful Dead fans that had cultural influence far greater than its membership numbers, in part by giving free accounts to journalists. Among bulletin-board communities of the time, it was rare in both its aspiration of achieving a viable business and the extent of its visibility in the popular press. On both counts, it served as a decisive bridge from the era of hobbyist online spaces to the commercial internet that Silicon Valley would produce.[8]

FIGURE 5.

Howard Rheingold subtitled his 1993 book of reportage on The WELL *Home-steading on the Electronic Frontier*. He did not initially develop the meaning of *homesteading* beyond the subtitle's implication that it described his newfound virtual homeland. In the book's 2000 edition, Rheingold refers to the term as "obsolete and anachronistic," a relic of a "pioneer culture" since lost to the internet's mass adoption and commercialization.[9] Yet Patricia Nelson Limerick has shown that the conquest of the western United States—the source of the "homesteading" and "frontier" metaphors on which Rheingold relies—is an "unbroken past" rather than a finite era that ended with a particular milestone of warfare or railroad construction. The frontier imaginary similarly persists online. As recently as the mid-2010s, the names of the first two major versions of the blockchain protocol Ethereum were Frontier and Homestead.[10] Rheingold's metaphors, then, have survived long after he and his fellow pioneers set out to explore, name, and demarcate the virgin territory of the "Net." Eventual commercialization was not the end of this process but its purpose all along—in digital space as much as on Indigenous lands.

"Western American history," writes Limerick, "was an effort to draw lines dividing the West into manageable units of property and then to persuade people to treat those lines with respect."[11] Homesteading became enshrined in US law with the first Homestead Act in 1862. It was wartime legislation, seeking the expansion of "free labor" against Confederate slavery, inviting Northern White settlers to populate Western territories based on made-up allotments of land deemed the appropriate size for nuclear families. Whereas Iberian dominions in the Americas parceled out land in large chunks to aristocrats, leaving subsequent inhabitants to demand disruptive waves of land reform, the homestead doctrine was to be a parceling-out of democratic ownership—democratic in the sense of personal, private, and widely available, but with a feudalism inscribed inside. Within the homestead, the male citizen was sovereign over his family, and through his dominion he became a democratic subject on his visits to town. Democracy thereby depended on the dual subjugation of the household and of the people whose territories preexisted its property lines. Part of the price of those homesteaded plots was the armed settlers' participation in denying existence to the Native peoples, for whom landowning was a foreign logic and whose livelihoods were often incompatible with the imposition of fences.

The homestead turns land into a bounded political object, encoding participants as the citizens who could be the basis of new states for the Union—although the land was not by custom or morality the US government's to give. Homesteading extended the earlier "doctrine of discovery," a theological-political principle that Christian settlers could assert title over non-Christian lands they conquered. Motivating settlement to expand the new United States required the mobilization of Evangelical Christian concepts like conversion and mission.[12] The thrall of democracy became a political gospel, calling the land into service and a new ethno-state into being.

Early internet products such as GeoCities and eWorld relied on metaphors of terrestrial and spiritual conquest to introduce their brands to customers still skeptical about online services. Digging a well—as in The WELL—was often necessary for permanent settlement and agriculture. Californian tech "evangelists" have aided startups in overcoming their initial nonexistence, asserting their impending reality with such confidence as to summon the necessary multisided markets and network effects. For these platform barkers, too, the promise of democratizing access to the wonders of software is at the heart of the product pitch.

Barbrook and Cameron devoted considerable exegesis to the Californian aspiration of "Jeffersonian democracy"—a utopia that they predicted would produce a dystopia of "cyborg masters and robot slaves."[13] Their prediction was that history would repeat itself. A condition of possibility for American homesteading was Thomas Jefferson's Louisiana Purchase, the acquisition of a French land claim that became roughly the middle third of the contiguous United States. This land, for Jefferson, would be the basis for a democracy of landowners—those feudal lords in microcosm—whose political rights derived from their local absolutism, just as his statesmanship depended on the labor of people he regarded as his slaves. Similarly, the design of social software exhibits that paradoxical politics: democracy is supposed to somehow emanate from the feudal. As in the homestead, the two tendencies are enmeshed and codependent, despite their contradictions. Democracy is the goal, even if it is not recognizable in the means.

According to the design pattern of implicit feudalism, nearly all social-media software nudges users toward autocratic or oligarchic forms of community governance, lacking the means for even the most typical structures of associational life offline. Punishment for wrongdoing is censorship of one's posts or exile from a given jurisdiction. The encoding of implicit feudalism into social software does not outright determine users' behavior, but it does bear a kind of politics, just as homesteading encoded the politics of property and patriarchy on its land claims. Whether the servers sit in an office closet in the Sausalito houseboat district, like the Rheingold-era WELL, or among corporate data centers around the world, the structures of power take cues from their technological substrates.

The shortest, least specific of The WELL's "design goals" stated, "It would be self-governing . . . ."[14] But the ellipsis never quite resolved. Rheingold later wrote, "Technically, the early WELL was governed as a benevolent dictatorship."[15] It obtained early members from the dissolution of The Farm, a famous counterculture commune in Tennessee that began under the rule of its spiritual leader, Stephen Gaskin.[16] Farm veterans became The WELL's admins. Beneath them was a mélange of group-level, micro-dictator "hosts" and seemingly endless, structureless discussions referred to as "meta." In 1994, the platform was sold to a new owner; users had no say in the matter. The buyer, the shoe magnate Bruce Katz, attempted to ingratiate himself to his newly acquired community with what could serve as a pithy summary of the Californian ideology: "I believe in the power of this new emerging media and believe that it is one of the bright hopes that we have in reinvigorating a civil dialogue that is the foundation of a free democratic society."[17]

In search of real self-governing, Rheingold and other WELL dwellers later formed The River, an online community owned by a cooperative of its users. But it never flourished. The WELL itself was acquired by a group of users in 2012, opening the door for self-governance only after the heyday of its influence.[18]

Compare the homesteading tradition to another sort of home, the "site of resistance" that bell hooks has celebrated as a *homeplace*. She explains: "Black women resisted by making homes where all black people could strive to be subjects, not objects, where we could be affirmed in our minds and hearts."[19]

Those who could not leave an oppressive society could find liberation together, transforming space and time, however constrained the homeplace might be by the world outside. The homeplace forms a counter-tradition to the homestead, a place of care and resistance, where power can be shared in contrast to the domination of the broader society in which it occurs and from which it can never fully depart.

There are elements of the homeplace in many online spaces, in what people have made with the Californian ideology's products, constructing sites of resistance again and again, beyond the knowledge or comprehension of the technologists and executives. Homeplaces have become particularly important among marginalized groups, whose members can find each other online in ways unavailable before. Tech companies have celebrated when social movements arise on their platforms, but those movements are not theirs.[20] Solidarity forms through the affective affinities among participants, regardless of who is technically in charge of the platform or the forum. The intimacy, the care, the rebellion, the imagination—none are in the code, but homeplaces occur both because of and despite the designs of homesteading machines. The feudal power flows are never the whole story.

Homeplaces came and went on The WELL. But contra Rheingold, the homesteading didn't end when communities moved to corporate servers. Digital space is an ever-expanding sort of West; the land is as limitless as server capacity allows, and the enabling factories and rare-earth mines can remain far from view. Within each pocket of delineated social space, what virtual terrain a user claims becomes their castle. If you don't like it, you can always find another plot to call your own. On a group chat, leaving is only a button away. As the libertarian political philosopher Robert Nozick wrote, the only utopia is the ability to exit one utopia for another.[21]

Exit has assumed an exalted place in Californian thinking. The availability of exit became the implicit justification of implicit feudalism: if a community is exit-able, that is enough to call it democratic. At the level of business, exit is the goal investors expect their startups to aspire to, in the form of an acquisition or public stock offering.[22] At the level of culture, the annual Burning Man festival practices the art of temporary co-creation and departure. Elon Musk opposes unionization in his terrestrial factories, but once his companies make possible the exit of Mars colonization, he hopes to establish "direct democracy" there. Upon acquiring Twitter in 2022, faux-democratic performances became part of his dictatorial management style; he claimed he would abide by the outcomes of user polls on company policies, despite employees' warnings that Twitter polls were insecure and vulnerable to manipulation. From dreams of space travel to floating "seasteading" colonies in international waters, the Californian ideology longs for homestead archipelagos, where feudal governance can finally flourish—justified by exit options, rebranded as democracy.[23]

Alongside the option to exit in the Californian imagination is the dream of scale.[24] Scale became an economic necessity. Silicon Valley's rise as a stronghold of consumer technology was a response to reductions in public investment through defense contracts. Early computer companies scrambled to develop an alternative source of money for expensive innovation, and they found one: venture capital, an investment strategy based on risky companies capable of dominating entire markets, so that the winners can pay for the far more plentiful losers. In 1979, VC investors got both a tax cut on their profits and a change to the federal "prudent man rule," enabling big pension funds to pour billions of dollars into these deals.[25] VC relies on business models seeking to achieve monopoly-level scale with near-zero-marginal-cost software. Implicit feudalism provided a social and technical blueprint to help founders and VCs maintain centralized control even across vast digital empires.

Politics can be slow, and its sensitivity to context interferes with limitless growth. Homesteading with implicitly feudal systems presented a way to bypass politics and keep scaling. If a particular entrepreneurial fiefdom doesn't work out, members can always exit, start another, and keep the network expanding. What Californian investors demand is clear: keep growing, consuming, colonizing, replacing—or cease to exist.

The Californian ideology's politics of no-politics encoded a social order into its tools and their surrounding institutions: the feudal permission-control logics of the technology at hand and the historical habits of homesteading. Barbrook and Cameron predicted the endgame as, rather than marvelous connection, "a deepening of social segregation."[26] Elite access to artificial intelligence and medical wonders would enable salvation by escape, a faithless religion of exit. From the comparatively minuscule WELL to Instagram, homesteading spread through the organizing patterns of daily life in digital spaces. Homeplaces may blip in and out of existence. But under the guise of an aspiration to "be self-governing," the more rigid powers of admins and CEOs alike are hard-coded to outlast the homeplaces. As Barbrook and Cameron suspected, this ideology would spread far beyond the platforms themselves, into mass politics.

A FEUDAL UNIVERSE

Soteriology is the branch of theology that deals with salvation, with whatever it is human beings should ultimately be striving for. A classic example is Anselm of Canterbury's eleventh-century treatise *Cur Deus Homo*, a feat of especially explicit feudalism. His account of a person's relationship to God extrapolates from the dominant political relationship of Anselm's eleventh-century world: subject and lord.[27] The relation is that of perfect hierarchy. God became human in Christ in order to make the only sacrifice worthy of the ultimate Lord. Anselm wrote the book as archbishop of Canterbury, a position that would have put him in frequent contact with the top of the feudal power structure, and that made him responsible for justifying the structure to its massive underclass. His local politics translated into his cosmic order. To be saved is to inhabit that order fully. The spiritual and political orders co-create each other.

The Californian ideology has a soteriology of its own. Barbrook and Cameron describe the Californian endgame as the parallel dreams of an "electronic marketplace" and an "electronic agora":[28] a frictionless economy and limitless speech that, if society accepts them, would wipe away the troubles of the analog world in a flood of true democracy. The flows of online life, that is, were to be vehicles for a kind of bloodless revolution, a salvation that investors could get richer by enabling.

A decade into the twenty-first century, the democratic prospects of social networks seemed real, especially as networked activists organized movements that unseated dictators in the Arab world and took on financial elites. But even then, the technical logic of implicit feudalism was shaping perceptions of the movements' politics.

One catalyst of the 2011 Arab Spring protests was the Facebook page "We Are All Khaled Said," created and controlled by Egyptian Google employee Wael Ghonim.[29] For Ghonim's role as the page's founder, the world press declared him the leader of the Egyptian uprising, although he lived outside the country and continually insisted that the movement was "leaderless." Later that year in the United States, the Occupy Wall Street protests exhibited similar contradictions. Veteran news anchor Dan Rather identified activist Priscilla Grim as "the real leader of this movement" because she happened to administer key social media accounts— a perplexing claim for a movement whose insiders, like those in Egypt, stressed their leaderlessness and used an offline, consensus-based assembly to make decisions.[30] Online activism was indeed instrumental for these movements, but the power structure of social media seemed to speak louder than the power structure articulated by activists themselves. In the streets and squares, activists were organizing through radically democratic processes, seeking to elevate direct participation over the representative systems that they denounced. But outsiders defaulted to the feudal logic of the protests' online spaces, assuming that technical workers were also movement leaders.

Before long, feudal systems gave rise to even more disruptive forms of feudal politics. The new religious movements are revealing. From the civil war following Syria's 2011 protests, combined with the failures of US-backed regime change in Iraq, came the Islamic State. It was not a Westphalian nation-state but a networked *umma*, a transnational community operating through the opt-in membership of hashtags and the imposition of absolutist order in its domains. As the Islamic State idea spread through brutal, viral videos and social-media groups, the Californian ideology's anything-goes social liberalism did not take hold. But the homesteading did—in this case adapted to the frontier of a stateless war zone, an act of exit from the international order. The implicit feudalism of the networks decoded there into an archipelago of territorial feudalism.

Horrific spectacle has been only one side of the Islamic State's media output. Rather, as Marwan Kraidy points out, "a majority of official I.S. visual media releases focus on non-violent aspects of life in the Caliphate": "in terms of a socio-religious utopia, it articulated claims of a pure, authentic, and truly Islamic society unburdened by Western influence and local subversion, with images of the good life—premised on a puritanical vision of Sunni Islam—showcasing spectacular sunsets and Ferris wheels and showing contented-looking people—mostly men—shopping in markets, fishing in rivers, praying piously, conversing amicably."[31]

These were the images of inhabiting a salvific order, with a clerical sysadmin. The implicit feudalism of the network expressed itself in an organizational hierarchy. The founding caliph, Abu Bakr al-Baghdadi, appeared publicly only in choreographed events designed for viral circulation, such as his 2014 proclamation of his alleged caliphate at the al-Nuri Mosque in Mosul, Iraq. The rest of the time, under his hegemonic absence, his ultimately fleeting regime portrayed itself with a virtual reality of ordinary life.

Meanwhile in the United States, the favored political party of Silicon Valley lost to Donald J. Trump, who turned Californian tools into his political home for movement building and then for governing. Alongside his presidency came the QAnon movement, a kind of digital gnosticism that blended Trumpism with Evangelical Christianity.[32] It produced devoted followers of a pseudonymous prophet, a government official named Q, who prophesied a salvific restoration of American society through a military coup and mass executions of the president's enemies. Trump's continued and unobstructed power would be assured. Before long, sympathizers won seats in Congress.

In the documentary *Q: Into the Storm*, director Cullen Hoback meanders to the conclusion that the author of Q's "drops" is Ron Watkins, the system administrator of 8chan, a website where Q posted. The same person rushing to get the servers back up during an outage, Hoback begins to suspect, also masterminded the apocalyptic movement. At critical instances, Q seems to have inside knowledge of the servers' workings. Watkins claimed to be in contact with the Trump White House surrounding the contested 2020 election; his powers as an admin brought him to the brink of participating in a political power grab. At the end of the film he seems to give up the disguise altogether, all but admitting to his dual role—a conjunction that further linguistic analysis has corroborated.[33]

Along with the CEOs of corporate social media who de-platformed Donald Trump in the last days of his presidency, Watkins represented a turning point. Earlier in the life of the Californian system, admins merely maintained the allegedly neutral platforms.[34] But now that story was giving way to regimes of platform diktat, handing all power to the admins. Trump soon created a social network of his own. Starting with the feudal designs encoded into their systems, the minutiae of technical administration expanded to become coterminous with geopolitics.

Watkins does not appear to have had a specific policy agenda to promulgate; he performs the studied indifference of online trolling culture.[35] During Trump's reelection campaign, similarly, the Republican Party broke with past practice and did not issue a policy platform. The Californian politics of no-politics had taken hold, through a grasp on power—server power, executive power—that could operate on its own terms, not in service to any external commitments. The salvific promise of Q was to overcome democracy and install the order of a platform homestead in its place. As with the Islamic State, the movement born on decentralized networks adopted the organizational default that implicit feudalism promulgates.

Perhaps no one exemplifies the actual soteriology of the Californian ideology like Curtis Yarvin. A blogger and tech entrepreneur, Yarvin has had the audacity to apply the commonplace structure of startup companies to politics. The result is outright, explicit monarchism—along with racism only lightly disguised in dog whistles. Yarvin's benefactor has been the influential Silicon Valley investor Peter Thiel, who was also an outspoken supporter of Trump's 2016 presidential campaign. Trump advisor Steve Bannon has been a Yarvin reader, and Yarvin was reportedly in communication with the Trump White House.[36] But political trysts aside, the basic alignment was to be expected: a coup-inclined president who came to power by tweeting, a tech industry organized through monopoly power, and a technologist willing to dispense with the pious fiction that his industry's achievements somehow incline toward democracy.

Howard Rheingold had seen danger in online social media back in the early 1990s. "Whoever gains the political edge on this technology will be able to use the technology to consolidate power," he wrote.[37] Ephemeral bursts of protest continue to spread across networks, and some of these call for democracy still. But the most novel, persistent kinds of spiritual-political imaginaries that have arisen on Californian tools are teaching more feudal kinds of lessons, a salvation that comes from ceding all power to the sysadmin.

EVERYDAY FRACTALS

Writer and activist adrienne maree brown recalls posting, in March 2016, an invitation on Instagram: "I am inviting a small crew of women and gender nonconforming friends into an experiment with each other, to share daily portraits of ourselves in this private thread for a month as a liberation technology, and affirm each other's beauty. Interested?"[38]

Six people responded and joined her online homeplace. "What emerged," brown wrote a year later, "was a community, a safe space, that is still very active today." Her recollection, with glimpses of what ensued, comes in her guidebook for social-change movements, *Emergent Strategy*. Rather than offering grand

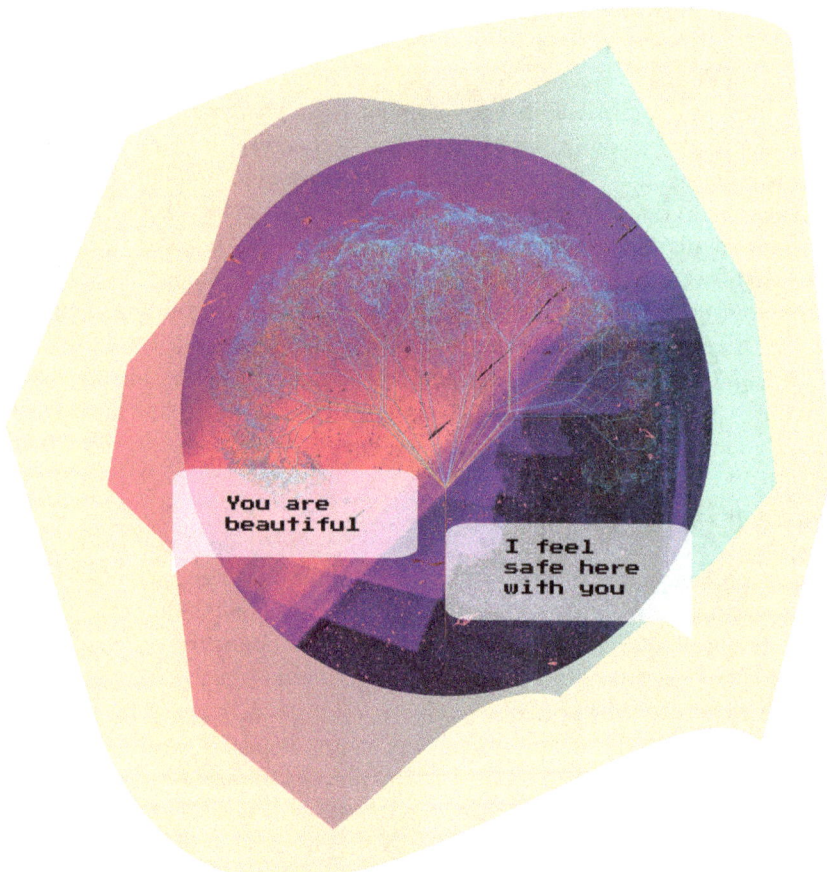

FIGURE 6.

strategies of conflict and policy demands, brown dwells in what Michel de Certeau called the "tactics" of everyday life.[39] She has been director of an important environmental justice organization, but readers looking for tips on institutional design and policy advocacy find instead the minutiae of intimate communities, along with a spirituality she draws from the novels of Octavia Butler and the pop-science of fungi and fractals.

Through this outlook, brown diagnoses the state of US democracy through the texture and practices of the everyday:

> We—Americans—don't know how to do democracy. We don't know how to make decisions together, how to create generative compromises, how to advance policies that center justice. Most of our movements are reduced to advancing false solutions, things we can get corporate or governmental agreement on, which don't actually get

us where we need to be. It was and is devastatingly clear to me that until we have some sense of how to live our solutions locally, we won't be successful at implementing a just governance system regionally, nationally, or globally.[40]

The everyday, then, becomes for brown the fundamental point of departure for social activists. "When we speak of systemic change, we need to be fractal," she writes. "Fractals—a way to speak of the patterns we see—move from the micro to macro level."

While Barbrook and Cameron placed the Californian ideology at the register of political economy, I have argued that Californian politics also reverberate in users' everyday experience with products. The everyday can be a site of enchantment, as for the Jesuit priest de Certeau, or of disenchantment, as when Henri Lefebvre details the deceptions in the life of a country church.[41] Ben Highmore summarizes de Certeau and his ilk like this: "What would a politics be like that emerged from the everyday, instead of one that was simply applied to the everyday?"[42]

Philip E. Agre was a precocious engineer and then a humanities professor before he abandoned academia for intentional obscurity in 2009. He is now credited with having predicted the looming regime of online surveillance—back when the Californian ideology feigned innocence about anything of the sort.[43] Like brown, he became fascinated by fractals and the relationship between the everyday and the world-historical, the minute and the immense.

Agre's dissertation at the MIT Artificial Intelligence Laboratory was called "The Dynamic Structure of Everyday Life." It includes an eighteen-page analysis of "walking to the subway," which serves to justify a shift in software design from the intentional to the improvisational. In the dissertation, as well as in a talk on the structures of everyday life while still a student,[44] Agre proposed the mathematical concept of the lattice as a gateway between the particular and the general, the local and the global, the routine and the complex. His lattice functions much like brown's fractals.

Almost two decades later, Agre returned to the lattice in an essay on political theory, alongside fractals and another long-standing keyword of his: *skills*.[45] Across his lattice structure, four dimensions of political skill form a network of intersections that cascade across society: vertical (from national to international), geographic (from local to global), institutional (from one institution to many), and ideological (from one commitment to networks of commitments). Along each dimension, skills that people develop in practice at small scales extend across larger scales of political life. A healthy society requires people exercising skills on all these dimensions. "The issue lattice is sufficiently complex," Agre writes, "that it will never emerge without high levels of political skill diffused throughout the society." While mass media and civics classes teach politics in terms of vaunted officeholders and halls of power, he held that lived politics depends much more on moving skillfully among the lattices.

Skills for Agre are both practical and mystical, a reorientation of all meaning-making as emanations from small acts of community. The epigraph of the book based on his dissertation is a medieval Zen dialogue. It begins:

> Joshu asked Nansen: "What is the path?"
> Nansen said: "Everyday life is the path."
> Joshu asked: "Can it be studied?"
> Nansen said: "If you try to study, you will be far from it."[46]

If there is a theory of salvation here, it comes through the friction of involvement, not electronic optimization. Technology must support the work of human politics, not replace it. Rather than flame wars, technologies might thereby encourage the art of consensus making, as brown teaches in *Emergent Strategy*. They might enable movements to persist and evolve, rather than disappearing into the next viral moment. If the dream of the Californian ideology is a world without politics, however, it stands to reason that the technology it generates would not teach political skills.

As the Californian ideology's anti-politics established itself on the West Coast, Agre was inverting it at MIT, calling for technology that invites people into developing skills through everyday politics. Agre concludes his essay "The Practical Republic"—the final essay listed on his faculty website before his sudden departure from public life—like this: "Technology is not central; what is central are the choices that we make, each of us, in laying claim to the rights and responsibilities of citizenship in our own lives."[47]

The technology that we need is technology that does not take or demand credit. brown seems to forget about Instagram upon summoning her community there; the homeplace becomes the subject.

What else could developing political skills look like? Perhaps it is a classroom where students collectively decide how best to play the game SimCity—taking a master-of-the-universe interface design and adding to it an exercise in face-to-face democracy; perhaps it is thousands of people collectively deciding on actions in a live-streaming game.[48] It might look like those occupations of public spaces during the protests around the world in 2011, when activists learned and practiced consensus processes with masses of strangers at once, experimenting with a kind of democracy beyond the elected officials and corporate boards that they believed had failed them. People learned new hand signals and techniques of persuasion, how to facilitate an effective meeting and how to disrupt one if they needed to. Occupy Wall Street developed a website where participants could keep track of the schedule of assemblies and the text of proposals that would be discussed. Occupy activists in Wellington, New Zealand, encoded their governance practices into an app, Loomio, that has since been adopted by organizations and even governments far from their island.[49] Although Loomio began by mimicking Occupy-style processes, it has come to support a wide variety of techniques for coming to agreement. Users can rank choices in order of preference, for instance, or invite volunteers to see who will actually implement a decision.

The platforms born in protest did not take their designs from a business model or technical convenience so much as from what their users were already doing offline. For Agre, likewise, cultivating political skills should precede the making of technology to support those skills. Recall Conway's Law, the notion that the designs of technical systems end up resembling the organizations that design them; to build anti-feudal systems, Agre would likewise stress the need to start by practicing anti-feudal interactions, wherever we find ourselves.

If a butterfly flapping its wings can cause a hurricane a world away, as the cliché goes, then anything could happen between a private thread on Instagram and a protest movement. When brown and Agre wield their fractals and lattices, they do so not with a comprehensive account of the causality. "Being a part of movements is complex work," brown writes. "It requires a faith."[50] With this kind of faith, and with everyday skills, brown and Agre reject the ultimate exit of Californian ambitions: the departure from bodily limits and social constraints. They refuse to regard technology as the angel of history, the divine agent, and instead insist that we are still just talking about how people relate to one another. Against feudal technology and the authoritarian revival it helped produce, the retort is not another technology, but the practice of political skills. If we honor those skills, perhaps designers will encode future technologies that nourish, rather than evade, everyday politics.

SUPERHIGHWAYS

In a bittersweet afterword to *The Virtual Community*, Howard Rheingold recounts how the intimate homeplaces he experienced had become a matter of industrial policy. Al Gore, first as a US senator and then as vice president, had promoted the "information superhighway" as a market and geopolitical opportunity.[51] It was protocol infrastructure that government would build and set free into the world. Rheingold noted the derision that the "superhighway" moniker had attracted—hyperbole compared to the experience one had on dial-up modems in those days, although faint in comparison to the homesteads the internet would soon bring. Yet as democratic skills erode through the everyday politics of online life, confidence in the plausibility of democratic infrastructure has eroded too. This has opened an opportunity for everyday feudalism to deepen its influence on geopolitical imaginaries.

When I took a high-speed train between Hangzhou and Shanghai, I sent a video home to my kids. No train like that exists where we live in the western United States. The legacy of homesteads here developed into a politics that made the assertion of public transit over private property too costly. The mightiest feats of infrastructure we drive by—the dams, the rail bridges across valleys, the tunnels through mountain passes—date to the 1930s or the early Cold War, the years when US president Franklin Roosevelt's "arsenal of democracy" was gaining strength.

Several years ago I had a long correspondence with a self-described Chinese student, who said she came to the United States to study and disagreed with something I had written in favor of democracy. She wrote: "China's achievements in human development are historically unprecedented. Under our system my generation has thrived, and is far more positive and forward-looking compared to our peers worldwide. There may be a 'perfect' model of democracy that you have in mind, but democracy as practised throughout most of history is best described as corrosive and sclerotic. One need only contrast the state of American and Chinese infrastructure to arrive at this conclusion." Two days later, she added: "Every inch of progress China had made resulted from an absolute, unequivocal rejection of democracy."[52]

Even those who claim the mantle of democracy appear to have come to similar conclusions. US platforms present themselves as the new arsenals of democracy; CEOs like Zuckerberg defend themselves against antitrust enforcement by arguing that their consolidated power is necessary to counter that of ascendant Chinese platforms.[53] This is a profound concession of democratic possibilities for the sake of expediency. But daily experiences with implicit feudalism in online life, as well as daily experience in countries whose democratic experiments have calcified, seem to insist on autocracy as an inevitability.

In China, autocratic order has a long history, always intertwined with social technology. For many centuries, emperors used technology to consolidate power— tracking the minutiae of production, exacting taxation—in ways European rulers could only dream of.[54] This order produced a discourse of "harmony," still a favorite word in Communist Party slogans.[55] Harmony is an article of Confucian faith, applied to assert cohesion against the lived experience of a society exploding into the overlapping complexities of markets, networks, and megacities. Autocrats aspire to produce the harmonious interplay of social roles among their subjects, though everyday harmony can eclipse even the autocracy. As Xiaobing Tang describes the outlook of writers in post-revolutionary China, "The emergent hegemony is no longer Ideology or Collectivity, but rather everyday life."[56] Ancient emperors ruled by their precision agronomy; now, implicitly feudal platforms and apps play that role. Under the fear of state crackdowns, the admins of Chinese social-media platforms and of their user communities act as subsidiary bureaucracies, protecting their right to exist by imposing their best guess of what harmony will allow.[57]

Benedict J. Tria Kerkvliet introduced the concept of "everyday politics" in the context of research among Southeast Asian peasant farmers. What he observed is also salient across the global diaspora of Californian technology: "Everyday politics involves people embracing, complying with, adjusting, and contesting norms and rules regarding authority over, production of, or allocation of resources and doing so in quiet, mundane, and subtle expressions and acts that are rarely organised or direct."[58]

What might look like the opposite of politics, that is, may be upholding or unraveling the reigning regime, a "power of the powerless"[59] in which ordinary actions can bear world-historical freight. The everyday can thus become a site of resistance.

"THERE ARE ALTERNATIVES"

This chapter has offered a rereading of the Californian ideology's politics of no-politics beyond the earlier focus on political economy—from the logics of homesteading and feudalism in ordinary online spaces to their role in enabling the rise of national authoritarianism. I suggest that everyday online practices at least partly tell the story of broader political shifts underway. However, I share Barbrook and Cameron's conviction that "there are alternatives."[60] Just as they point to the French state's Minitel system as an alternative political economy for networks, I find that alternatives lie in the everyday politics of hooks's homeplace, of brown's deliberate interdependence, and of Agre's political skills. Producing more democratic and humane politics at large scales requires attention to the daily political practices on networks, as well as to what software designs might encourage or discourage.

Politics is no autonomous category in human minds and worlds; for that reason I have sustained attention on the diverse forms of religious imagination that have aided the global decoding and re-encoding of Californian tools. Frontier evangelizing, apocalyptic Islam, Confucian harmony, and the faith that fills a homeplace all inscribe their meanings on network spaces. These imaginaries are a reminder that, along with political skills, the production of alternatives must involve dimensions of ritual, devotional commitments, and structures of belief.

Anselm of Canterbury lived in a certain kind of feudal world, a world whose daily interactions of power and deference informed his view of the spiritual order. An online world composed of implicitly feudal systems has similarly informed its inhabitants. Through daily practice, they have learned political skills more oriented toward fixed authority than democratic accountability. The skills one has are the skills one can imagine using. To cultivate different skills, therefore, is a task of not only technological design but also of imagination and spirit.

FIGURE 7.

A People's History of Twitter

betterplatform.net

When Elon Musk acquired the social-media platform Twitter for $44 billion in October 2022, it was a stark reminder to many people that our online civic spaces are commodities that can be bought and sold. Especially when Musk's early weeks came with scorched-earth layoffs and disorienting policy changes (including ones that targeted journalists), users began fleeing to other platforms.

Back in 2017, I was part of a team that created a shareholder proposal at Twitter, aiming to decommodify the company by establishing a framework for its users to become its owners. Five years later, after Musk's takeover, we began experimenting with a different strategy: imagining what it would be like if the platform had become a common good in service of the global public sphere. We teamed up with former Twitter workers to reflect on the kind of platform they had hoped to build at the company. We also learned from people who had led and studied citizen assemblies for governments. How might a representative assembly of Twitter users work? What kinds of proposals might it make? Who would have to agree to them if the users were in charge?

Before establishing some kind of shadow government to devise an alternative future of Twitter, which Musk has since renamed X, we decided to start by grounding ourselves in the past. We organized an online event in March 2023 called "A People's History of Twitter," which attracted nearly two hundred technologists, journalists, activists, and other users. We also released an online chatbot that people could use to share their experiences with Twitter over the years. This collective history provides a foundation for articulating expectations about what should come next—what people have loved about Twitter, and what they hated, what the company did right, and how it betrayed us. These questions matter, whether Twitter users decide to stay or go to another platform.

The People's History included accounts of people finding jobs, new friends, spouses, antiracist organizing, queer communities, fashion, and news about niche topics. They experienced Twitter as a place of self-expression, of finding a voice and an audience they didn't have before. Many also described having distanced themselves from it more recently. In that sense, the People's History served as a kind of wake, a joyful way of mourning something that, at least in some respects, had died.

Governance is not just about holding power and making decisions. Before a community can begin to self-govern, it needs to see itself as a community—through participants telling stories about themselves and having shared experiences. A People's History of Twitter was an attempt to begin that process, to initiate people's transition from being users of someone else's platform to being full citizens of the networks they live by.

Democratic Mediums

Case Studies in Political Imagination

So far I have offered a diagnosis: a story of how the design pattern of implicit feudalism and the inherited ideology of homesteading inhibit the exercise of self-governance in online social spaces. I further suggested that, as more political life moves online, a widespread lack of experience with sharing power has lent fresh appeal to authoritarian urges. As people understand themselves more through their identities as users of social media, they find little reason for faith in their capacity to self-govern with each other. When compounded on the scale of billions of people-turned-users, intimate online experiences can have world-historical consequences.

From here on, I shift from diagnosis to remedy. The remedies I explore involve rethinking the design and practice of online social spaces by treating democracy as itself a medium for struggle, play, and policy. This is a departure from more widespread calls for online platforms to better serve the legitimate functioning of representative elections among territorial governments.[1] While I do not object to such calls, I focus instead on remedies that match the ailment I have observed— remedies that bring democracy more deeply into everyday online life, establishing appropriate kinds of jurisdictions as sites of creative self-governance.

The cases in this chapter come from outside the kinds of institutions that typi-cally claim the mantle of democracy. This is to be expected. Dominant political and civil-society institutions habitually resist imaginative politics within their bounds, particularly when it means unlocking power for people who have long been excluded. Historical examples are plentiful. Permit me one about a remark-able scholar who happened to pass away during this writing.

On June 4, 1993, US president Bill Clinton withdrew his nomination of legal scholar and litigator Lani Guinier, who would have been the first Black assistant attorney general for civil rights. The move followed a bitter sequence of backroom deals and attacks in the legacy press, which she later called a "low-tech lynching."[2] The phrase echoes the accusation of a "high-tech lynching" leveled by Clarence Thomas on the Senate hearings that ultimately confirmed him to the Supreme Court. The treatment of Guinier was widely seen as retribution for the Thomas hearings; then-Senator Joe Biden played pivotal roles in both affairs, defending Thomas from accusations of sexual harassment and undermining Guinier. In a feat much like what the perpetrators would later decry as cancel culture, Clinton retreated before characterizations of Guinier as a "quota queen" and of her scholarship, in the words of columnist George F. Will, as "extreme, undemocratic, and anticonstitutional." The evening that Clinton withdrew her nomination, Guinier shuddered when she heard the president, a friend since law school, repeat the "undemocratic" part on television.[3]

The remark was particularly painful to Guinier in light of the fact that, as is apparent to any half-serious reader of her scholarship, advancing democracy was precisely her intent. To this end, she explored practical, tested alternatives to winner-take-all voting, such as proportional representation. "I expressed reservations about unfettered majority rule," she explained afterward, "to ensure fair representation for all substantial minorities."[4] Later, she described racialized minorities as "the miner's canary"—people whose experiences represent early warnings for social problems poised to affect everyone else.[5] Her academic proposals sought to imagine how existing institutions could better reflect their stated democratic values.

Ironically, many of Guinier's critics also built their political careers on the project of defending minority rights—only, in their case, the rights of already privileged minorities, through the defense of patriarchy and racial hierarchy. Calling Guinier's oeuvre "undemocratic" might have been just outright racism or political opportunism, but it also rested on a certain theory of democracy. The theory is shared among both the conservative and liberal branches of the US political establishment, as Clinton's deference suggests: democracy is coterminous with whatever existing institutions happen to be.

Institutions are a precious inheritance. Even flawed ones should not be discarded recklessly, at the risk of being left with something worse. Lasting traditions of thought and practice are indispensable to a thriving political culture.[6] Yet unlike feudalism and autocracy, democracy cannot survive in stasis. Alexis de Tocqueville's canonical view of "democratic revolution," a gradual progression toward ever more deeply democratic institutions, refuses to deify any particular institutional form. During his antebellum sojourn in the United States, he wrote to his father in France that "there is nothing absolute in the theoretical value of political institutions."[7] Tocqueville scholar Barbara Allen stresses that, for him, the anchor of democracy must be transcendent, beyond the scope of any possible government; the work of politics lies in "harmonizing earth and heaven," as he put it. Accounts of US civil religion typically associate divine favor with specific institutions. But the sort of democracy Tocqueville admired has its roots in a religion based on a covenant with the divine, one that "opposes any notion of absolute authority" on earth,[8] enabling a continual evolution of the power structure in practice. No institution can fully manifest human equality. Still, he observed how people could continue pursuing it through a shared belief in equality before heaven. Something is sacred, but it isn't any particular institution.

The patterns of democratic erosion worldwide, which this book opened with, suggest that centuries-old institutions are increasingly inadequate for confronting ascendant authoritarians. Merely defending a certain sort of democracy is no way to help democracy as an ideal. Permit me a cliché: the best defense is offense. Authoritarianism today takes different forms than it did in the past, adapting its tactics against those of democracy. For democracy to thrive, its institutions must be vulnerable to continual reinvention. Its traditions must be alive enough to permit that. The task of making online spaces governable, therefore, should begin with imaginations radical enough to transcend existing institutions, together with the playfulness to hone imagination in practice.

In what follows, I consider two very different cases of mediated democratic experimentation currently underway: transformative justice, a movement to abolish policing as we know it through participatory processes, and cryptoeconomics, a project of reimagining economics and governance through internet-native blockchain protocols. They are subcultures that do not typically see each other or speak with each other, and they rest on often-diverging sets of values. But I have been drawn to them both, despite and because of what distinguishes them. Both share the radical premise, notably, of attempting to organize self-governing infrastructures that do not rely on state violence to establish order. I present them as starting points for the reimagining the design of technologies as *democratic mediums*—mediums in the sense of both enabling self-governing communication and serving as meeting points for the transcendent and the everyday. These starting points offer insights for the making of truly governable online spaces.

Both cases reject the widespread preference for holding dominant quasi-democratic institutions as sacred, choosing instead the quest for more deeply accountable institutions and more lively encounters with tradition. They are both intensely contextual. They seek to design processes that are appropriate to the problems at hand. Both also expect people to gain and cultivate political skills, to be participants in crafting their own democratic futures. Their media production meanwhile reveals how they craft transcendent visions, which are both anchoring and always provisional.

For each case, I profile its political imagination followed by an analysis of lessons from its design practices for governable spaces. The chapter ends with a consideration of how imagination can take form in political play.

ABOLITION DEMOCRACY

When a wave of outrage surged across the United States in 2020, following the murder of George Floyd by a Minneapolis police officer, key organizers in the Black Lives Matter movement were ready with a demand: "Defund the police." Liberal politicians seemed poised to act on the call, at least until the street protests faded and an apparent crime wave moved them to reverse course. The call to "defund" received blame for Democratic Party defeats in the 2020 election, securing its loss of favor among liberal elites.[9] The advocates for defunding or outright abolishing police soon found themselves in a position not unlike that of Lani Guinier: to explore deeply the question of what might make society more democratic is to risk being labeled a traitor to democracy.

Underlying "defund" was a legacy of community-based organizing to address violence and conflict without policing, known as transformative justice. Although Black Lives Matter has often appeared in mass media through images of Black men facing police violence, many of the community leaders behind it identify as women, queer, or non-binary. Chicago-based activist Mariame Kaba became a mainstream voice for defunding with a *New York Times* opinion article, "Yes, We Mean Literally Abolish the Police"; for many years before that, she had been a leading participant and teacher in community accountability processes, aimed at addressing both interpersonal harm and its root causes.[10] These signify a struggle to address interpersonal harm with participatory processes among affected people that address root causes. "People like me who want to abolish prisons and police," Kaba wrote in the *Times*, "have a vision of a different society, built on cooperation instead of individualism, on mutual aid instead of self-preservation."

For Kaba and other Black Lives Matter leaders, transformative justice work had been all along embedded in their advocacy for the abolition of police and prisons. More than a fixed method, it is a practice of exploration among many paths toward safer communities. A seminal textbook summarizes this practice in its title, the *Creative Interventions Toolkit*. The authors, who formed a purposely temporary organization to produce the book, explain, "We call ourselves Creative Interventions because creativity is often just what is needed."[11] Kaba and collaborators have also developed a website cataloging abolitionist and transformative efforts, with a name reflecting that creative urge: One Million Experiments.[12]

Instead of referring incidents of harm in a community to the arbiters of state violence, transformative justice equips community members to build their own skills for facilitating conflict resolution and accountability. The goal is not to punish and coerce, like police and courts do, but to repair harm and enable people

Community
fridge

Library
for all

Transformative
justice center

Mental
health hub

FIGURE 8.

involved in it to establish healthier relationships. The call to "transform" also goes further: it means not merely resolving a given incident, but recognizing how wider injustices might have helped cause it.

For instance, while the legal system would respond to a case of partner abuse by charging one party or both with a crime and seeking to punish accordingly, a community accountability process would begin with conversations. How did each partner experience what happened? Along with a trusted facilitator and allies, they might meet in a circle, where the person who caused harm agrees to take responsibility for it and apologize. Forgiveness may or may not be involved. The process might further reveal that an unjust eviction had been exacerbating tensions in the relationship. Together, the participants develop a strategy for publicizing the landlord's behavior and making exploitative evictions less likely in their community.

A process like this is not straightforward; it requires art and skill. Kaba co-authored a workbook for advanced practitioners,[13] which includes techniques for running accountability processes alongside exercises to help practitioners reflect on their personal development. Eschewing the temptations of institutionalization, she reminds readers that she does not do community accountability work for hire.

"Safety is not a product that we can package and market," writes another transformative justice activist, Ejeris Dixon. "We are invited to practice community safety skills with one of our most precious resources, our lives."[14] A checklist in the *Creative Interventions Toolkit*, asking "Is This Model Right for You?" expects that readers affirm all of the following:

- Want to address, reduce, end or prevent a situation of violence (violence intervention)
- Seek solutions within your family, friend network, neighborhood, faith community, workplace or other community group, organization or institution
- Can think of at least one other person who may be able to work with you to address this situation
- Want to find a way to support people doing harm to recognize, end and be responsible for their violence (accountability) without giving them excuses (without colluding) and without denying their humanity (without demonizing)—if possible
- Are willing to work together with others in your community
- Are willing to work over a period of time to make sure that solutions stick (last a long time)[15]

Despite its centrality in the experience of prominent Black Lives Matter activists, transformative justice only rarely surfaced in popular narratives following the 2020 wave of protest. In most public discourse about the call to defund police—and this is in part attributable to the slogan itself—the focus was on the existing institutions of policing, rather than on what other institutional arrangements could replace it. Perhaps most importantly, far too few people had knowingly experienced alternatives to policing in their everyday lives; the experiments were still too contained, their stories too little told. Thus, when anxieties about crime rose after the protests—resulting at least in part from police withdrawing their labor—politicians could claim they had no recourse but to further fund the only institutional option available to them for reducing crime: the police.[16]

Since the choice appeared to be police or nothing, the advocates of defunding could be portrayed as being opposed to public safety, even a threat to it. Media narratives tolerated a debate only within the bounds of current institutions, not one about how institutional arrangements might be rethought. Many Americans did not support the call to defund police because they could not envision an alternative to policing, for both addressing crime and providing the many other kinds of social services that police have come to control.[17]

The writings of transformative justice practitioners tend to be far from the language of policy prescription. They assume an audience that inhabits the streets rather than the halls of power. Kaba called for moving police budgets to education and other basic needs in her *Times* article, but this is a literature that overwhelmingly prefers forms of exchange that nourish dialogical thinking and ongoing reinvention. Activists' books intersperse essays with interviews, suggesting that the movement privileges conversational thinking over dogmas. Kaba stresses that her workbook is not "a dictate or THE LAW."[18] Her co-author Shira Hassan describes the discovery that "always reinventing the wheel was a feature and not a bug of doing this work."

As if to hold the uncertainties, an assemblage of spiritualities accompanies the practicalities of this literature. Kaba and Hassan's workbook, *Fumbling Towards Repair*, has on its cover three people, connected with constellations, cultivating a flower that emits star-pollen into the sky. The text encourages introspective habits such as journaling and self-care. Activists speak of ancestral bonds, rituals, magic, breath, the natural world, and mystical cosmologies alongside political visions and practical tips for accountability work. Adrienne maree brown, a leading thinker on transformative justice, celebrates the religion of Octavia Butler's science fiction novels in which "God is change," a divinity standing against the temptation to place one's trust in a stable world.[19]

Together these gestures organize a shared rhetoric of "harmonizing earth and heaven," in Tocqueville's sense—a transcendent orientation for political imagination that can see beyond now-reigning institutions. But more than calling on a fixed referent like Tocqueville's Christian God, they draw on multiple reference points. Their transcendence is tied to community practice. Constellations, whether in the sky at sea or on the cover of a guidebook, help people locate themselves when what they see below the horizon is insufficiently trustworthy. Just as the Big Dipper led escaping slaves north in the Underground Railroad, *Fumbling Towards Repair*'s stars point toward a future at odds with the terrestrial institutions of the present. In a nod to that earlier struggle against the once-unshakable institution of chattel slavery, transformative justice activists often refer to their broader movement simply as abolition. But for them abolition is never merely a negation.

A godmother of the defund movement is the philosopher Angela Davis. Her experience with incarceration resulting from her activism in the 1970s established her as a leading abolitionist against police and prisons. In a later series of interviews, Davis echoes W. E. B. Du Bois's call for "abolition democracy."[20] For Du Bois, this meant rendering slavery finally obsolete by ensuring the place of former slaves in democratic institutions—voting rights, cooperative economic power, and access to education, for instance. For Davis, achieving abolition democracy means establishing the conditions in which police and prisons are no longer necessary, because more democratic practices have replaced them. Transformative justice is self-consciously a project of abolition democracy.

"Abolition is a fleshy and material presence of social life lived differently," writes Ruth Wilson Gilmore, another abolitionist godmother, paraphrasing Du Bois. Indigenous musician and scholar Leanne Betasamosake Simpson offers a further phrasing: "abolition unfolding."[21] But this more expansive view of abolition, the call to self-governance, rarely makes headlines.

A further challenge to the visibility of transformative justice is that activists have taken care to keep their accountability practices out of corporate-controlled online spaces. They stress that community accountability must inhabit a temporality distinct from that of social media, allowing processes to proceed at their own speed as opposed to fitting into the attention span of virality. They resist social media even though they practice it expertly; Kaba, brown, and others have large followings online and participate actively. Yet brown warns, "Real time is slower than social media time, where everything feels urgent."[22] Kaba's workbook has prefatory warnings about who should not use the book, which includes anyone "not planning to engage participants in person."[23] She elsewhere adds, in an interview: "I pretty much hate a lot of social media. I use it as a tool, but I'm not a fan of the way it can flatten people and can flatten issues, and sometimes allows people to remain anonymous in very harmful ways. That said, I've actually tried to think through with other people what are some potential guidelines that we might agree to, some rules of the road around engagement on social media if you're doing community accountability work and transformative justice work."[24]

Despite her reservations, Kaba thus recognizes that the kinds of processes she has tended cannot remain solely in-person forever. Community accountability is as much needed online as off because harm is happening online, and anyway the lines between the virtual and the real no longer hold. The virtual is also real. As far back as the famous case of sexual assault on LambdaMOO in the 1990s, discussed in chapter 1, the need for self-governing online often stems from the need to address and repair harm.[25] If online communities cannot self-govern, they cannot resolve conflicts as they see fit.

Online spaces need abolitionist imaginations. Intersecting experiences of oppression and marginalization run rampant there, along with the habits of punitive enforcement. Governments and technology companies offer to solve problems with rules and punishments, but those institutions represent the racial capitalism that abolitionists want to make obsolete with their own solutions. Kaba and Andrea J. Ritchie quote Grace Lee Boggs: "We need to exercise power, not take it."[26] Change of this sort cannot happen by replacing who is in charge, only by altering how power flows.

Boggs also frequently reminded her disciples to prioritize "critical connections" over "critical mass"—a conviction that the germ of seismic change lies in the thick relationality of how people choose to self-organize day to day, rather than in a mass of faceless participants.[27] This is not a retreat from large-scale social change but a reorientation to it. For her, abolition begins with a theory of society in which there is no need for "the masses" anymore, because the center of our attention has turned to people, their relationships, and their communities.

Subsidiarity, Scalability, and Accountability

What would it take to make online spaces work for community accountability? How might lessons from transformative justice begin to inform the design of social media? Amy Hasinoff and I have argued that doing so would require a shift in the design of dominant social-media platforms: a shift from *scalability* to *subsidiarity*.[28]

Anthropologist Anna Lowenhaupt Tsing defines the aspiration of scalability as "the ability to expand—and expand, and expand—without rethinking basic elements."[29] She warns that "scalability never fulfills its own promises." As the biologist J. B. S. Haldane wrote decades earlier in his whimsical essay "On Being the Right Size," "a large change in size inevitably carries with it a change in form."[30] Also between the world wars, while observing the advent of new mass-communication technologies, Walter Lippmann concluded that democracy could not translate from local town halls to the scale of large nations connected only by their broadcasts; in his view, control by a small elite would be inevitable.[31]

Scalability has become the business model for the venture-capital investment that underwrites nearly all corporate social media. The payoff for investing millions of dollars in an unproven startup is the prospect of a business that can add large numbers of new users at ever-declining per-user cost.[32] Platforms therefore seek to govern harm and conflict through software-enabled automation: global rule books, algorithmic enforcement wherever possible, and opaque human decision-making when necessary. The result is a regime that provokes continual complaints of both overreaction and underreaction to apparent bad behavior, born of blindness to context and lack of due process. Tarleton Gillespie has suggested, in sum, "Maybe we should not automate."[33]

This modest suggestion has far-reaching consequences. The less automated a system becomes, the less it can participate in the economics and design practices of scalability. But there is another way. Taking inspiration from transformative justice activists, Hasinoff and I show how participant-centered systems can adopt subsidiarity: a principle that prioritizes appropriately local control wherever possible, within a larger system.

Subsidiarity was first articulated in Calvinist and then Catholic theology—for instance, stressing the relative autonomy of each congregation or region in a wider church.[34] It has since been incorporated into secular politics, including the founding documents of the European Union. But well outside this Western lineage, the basic idea manifests in virtually any durable form of social order, from common-law judicial systems and bands within tribal nations to the distributed authority structure of Sunni Islam. When institutions lose context-sensitivity and local control, they risk being perceived as illegitimate. Residents of a newly established town may expect to have their own post office, library, schools, law enforcement, garbage collection, and elected council. That is subsidiarity. In comparison to almost any other kind of institution that claims to serve civic interactions, the faith in scalability among social-media companies appears peculiar.

I have seen subsidiarity at work with particular clarity in cooperative business. Near where I live, there are two large hardware stores a few blocks from each other. One is part of a national, investor-owned chain; it is like every other store of its kind, wherever you go. Help from employees is scarce because they are stretched as thinly as possible. The other store is locally owned but part of a national purchasing cooperative, a business owned by local stores, designed to make them more profitable. The co-op pushes value and control to the edges of the network—to the store owners—rather than accumulating both at the center, on behalf of distant investor-owners. That store is an anchor of our community. Helpful employees are everywhere. This is subsidiarity again: scale where necessary, such as in joint purchasing with other stores, but local control everywhere it matters.[35]

What transformative justice activists call for can be understood as a radical subsidiarity. They want to enable accountability not just at the level of cities or regions, but among neighborhoods and friend groups—a scale similar to that of many online communities. Accomplishing this requires the widespread cultivation of political skills so that people have the capacity to organize accountability processes wherever harm occurs. Online, this would mean that any community must have the tools and interfaces to develop processes that are right for its culture. Facilitators should have the tools to carefully manage a process. A process should not be exposed to the gusts of some algorithm's viral winds or to rules set in a distant corporate office. To be felt as legitimate, the process must be voluntary and sensitive to context, not imposed from above. According to one study of online moderation practices, "People's sense of being treated with dignity and respect appears to have the strongest correlation to overall fairness."[36] An experience of fairness lowers the likelihood of repeating bad behavior.

Subsidiarity, to be clear, is not a demand for limitless local autonomy. Alongside autonomy it involves relationships to larger systems. It is not scalability, but it does enable scale—composed of spaces small enough to be governable. While transformative justice happens primarily at the level of local communities, those communities do not act in isolation. Communities need to learn from each other, working in concert to transform systems beyond themselves. The root of *subsidiarity* is the Latin word for help—meaning the mutual help among communities that constitute a larger whole.

Subsidiarity is not everywhere alike. It can be federalist (with smaller units nested within bigger ones) or polycentric (with smaller units connected laterally across a network).[37] Both forms already appear in online spaces to an extent. The office-oriented chat platform Slack, for instance, is more federalist. A particular employee might manage a small "channel" within a "workspace" controlled by the company where they work, hosted on servers owned and managed by Slack, which is in turn a subsidiary of Salesforce. Implicit feudalism and corporate ownership limit the capacity for participant self-governance at each of those levels, but some degree of local control is real. In contrast, email is a more polycentric network, enabling communication across multiple servers that might each be governed differently. Another example of a polycentric network is Mastodon, an open-source microblogging platform that users can host on their own servers, while connecting to users on other servers. Even in the absence of a central company or other enforcer, Mastodon communities have shown the capacity to carry out large-scale enforcement actions against incursions from the Islamic State and Gab, a Mastodon-based platform friendly to White supremacists.[38]

Subsidiarity involves the capacity to hold communities themselves account-able, not just their members. Transformative justice activists warn against con-sidering "community" an unmitigated good; harm often occurs because of, not just despite, its host community.[39] The purpose of a community can be precisely to support harmful behavior. The call for transformative justice, again, does not mean transforming only individuals' relationships but their social contexts when necessary. For instance, the documentary film *Hollow Water* depicts an account-ability process in an Ojibway village where sexual abuse had become endemic. Carrying out a transformative process required pressure on the village from both the federalist Canadian legal system and the polycentric networks among fel-low First Nations communities.[40] Designing governable spaces should similarly involve not just self-governance within a particular community but accountability among communities, across networks.

Any attempt to apply lessons from transformative justice must take seriously its practitioners' skepticism about bringing their practices online. But if the affordances of social media they recoil against are bound up in scalability—the inhuman pace, the context collapse, the lack of community control—perhaps online spaces crafted with strenuous subsidiarity could be more amenable to context-sensitive accountability processes. This cannot occur through a quick, superficial fix to the user interface. Subsidiarity requires duplication and customization of systems at a local level—exactly what investors want to avoid paying for in their pursuit of scalability and market dominance. Community accountability is friction from the perspective of an investor's profit margins. It all looks like costs: training facilitators, supporting diverse power structures, and imposing limits on external control.

An abolitionist orientation sees things differently. Scalability, enforced through coercion, is no basis for real problem solving. Societies and networks where people can govern themselves are places where the ever-expanding cost of policing and punishment is no longer the only option available. The resources once spent shoring up a dehumanizing system can go toward investing in the people who will participate in an abolition democracy. As W. E. B. Du Bois recognized in the wake of chattel slavery, abolition is not complete unless it comes with the rights and the skills to co-govern.

Designing networks with deep subsidiarity may be possible only through regimes that reorganize the flows of platform ownership, so that power ultimately lies with users themselves. The next case arises from subcultures very different from those of transformative justice activists. But it presents an opportunity for reorganizing flows of control and value that online accountability processes could build on. It could be a way out of scalability. Getting there, once again, involves not just a technical or economic feat but leaps of imagination.

CRYPTOECONOMICS AND POLITICS

Moloch, an ancient Levantine god whom the Hebrew Bible accuses of abetting child sacrifice, has found a new cult. It began with a blog post by Scott Alexander, "a psychiatrist on the US West Coast": an interpretation of the Moloch portions in Allen Ginsberg's midrashic poem "Howl."[41] Alexander recasts Moloch as representing the soul-crushing establishmentarian systems that plague us only because no better means of coordination exists for replacing them. To him, passages from Ginsberg like this are actually about breakdowns of signal and shared intent:

> Moloch the incomprehensible prison! Moloch the crossbone soulless jailhouse and Congress of sorrows! Moloch whose buildings are judgment! Moloch the vast stone of war! Moloch the stunned governments!

> Moloch whose mind is pure machinery! Moloch whose blood is running money! Moloch whose fingers are ten armies! Moloch whose breast is a cannibal dynamo! Moloch whose ear is a smoking tomb!

Gone is the standard interpretation of "Howl" as a retort to technocratic capitalism. Gone also is Karl Marx's reading of Moloch as capital's claim over "all surplus-labour which the human race can ever perform" and money, "to whom everything must be sacrificed."[42] To Alexander the trouble is inadequate technology. According to his post, "Every single citizen hates the system, but for lack of a good coordination mechanism it endures"; in turn, "technology has the potential to seriously improve coordination efforts."

Alexander's heterodox reading of Ginsberg has since spread among certain clusters of entrepreneurs and engineers building blockchain-based technologies. In the headlines, blockchains have become widely associated with fraudulent pseudo-banks and spectacular meltdowns, with libertarian ideologues hiding assets away from regulators' reach. But in certain subcultures of entrepreneurs, the longer story of what they are doing is transforming the social order through new mechanisms for coordination.

Five years after its publication, the blog post's exegesis took financial form with MolochDAO, a software contract on the Ethereum blockchain devoted to the slaying of this new Moloch: the "god of coordination failure, who consumes our future potential for perverse immediate gain."[43] As a DAO, or decentralized autonomous organization, it is a creature composed of contracts written in computer code rather than legalese, funneling human inputs through its software. MolochDAO was meant to help move the world toward a new economic infrastructure in which networks and code, rather than police and armies, would be the basis of social order. In its more mundane practice, MolochDAO is a collective grant fund, a pool of digital money that participants contribute to and then allocate to projects they deem worthy. It was first "summoned" at the ETHDenver conference in 2019 by entrepreneur Ameen Soleimani. Appearing during a market downturn, its purpose was to provide funding for people to keep experimenting and building, along with a galvanizing mythology. The MolochDAO website invites visitors into a cosmic confrontation: "This demon god of coordination failure, who consumes our future potential for perverse immediate gain, will be slain. Pledge your oath to his demise, or go down with him."[44]

Concocting elaborate mythologies is common in the cultures surrounding blockchain technology. The original cryptocurrency, Bitcoin, began in 2009 with a Promethean "Genesis Block" and then adopted the idiom of metallurgy. Among its central technical concepts are "mining" and "minting," for instance, and then there is "wallet" to describe whatever medium happens to record the gibberish strings of characters that provide access to a user's holdings. While Bitcoiners implore each other to hold their tokens during a sudden price decline with the call to "HODL," the Moloch slayers remind each other to "BUIDL," to keep building useful things. Among DAOs, language helps obscure the degree to which apparent novelty is reproducing much more familiar organizational patterns. Proxy voters are now "stewards" practicing "liquid democracy." The banal quarters of corporate temporality—Q1 to Q4—become the more ecological "seasons." Committees are "pods." Clubs and collectives are "guilds" and "covens." Initiating a vote means issuing a "spell." Shares, money, and multitudinous other financial instruments can be programmed into digital "tokens." For some, part of the appeal of the DAO idea itself is the name's affinity with Daoism, as if that ancient philosophy were being rediscovered in code.[45]

It would be too convenient to dismiss what is going on as mere disguise or wholesale recapitulation. A new name invites the breaking of old norms or at least fresh iteration with them. Names matter. Tokens can resemble preceding financial instruments, but they nearly always break those molds in some way; the old distinctions between money and equity, or labor and capital, are not so clear and seem to be evolving toward new sorts of distinctions.

The new names enable participants to tell each other, at least, that the strictures of securities law and labor law no longer apply, inviting fresh abuses and innovations. But not all renaming is so cynical. Calling proxy voters stewards has occasioned new kinds of Web interfaces for evaluating stewards' behavior. The one-share-one-vote norm of corporations is meanwhile beginning to wane in the imaginative universe of crypto in favor of algorithms that balance a voter's stake with dimensions such as temporal commitment or the number of other voters;

these forms of tallying preference have little precedent in corporate governance.[46] Since crypto projects tend to avoid reliance on state identity systems, the meaning and basis of identity becomes an open question, both social and technical.[47] And while a season might correspond temporally with a quarter, the language evokes a blurry duration that, through its ecological and ritual associations, seems to alter the flow of time.

With crypto-tokens come distinct experiences of organizational belonging and internet browsing alike. Rather than logging into websites with a username and password, one shows them what tokens are in one's wallet with a browser plugin. One's identity lives in that wallet, not in the database of the website. With certain tokens might come the ability to trade or to vote on a proposal. Debates about proposals live across various chat threads and online forums, depending on the community—largely out in the open, in public, rather than in a closed boardroom. If there is a vote, it is likely on a futuristic website, in hacker-friendly dark mode, meeting the cultural habits of the early adopters where they are. When ownership and governance are a matter of points and clicks, they are intuitive and expected for users who live their lives in apps. They become part of flows in daily life like stock certificates and Bloomberg terminals never were.

To reimagine what is basically a giving circle as slaying Moloch through coordination does something to the nature of the giving. (Giving in MolochDAO is called making "tributes.") The point becomes less the gift than the art of orchestration that made it happen, the coming together of the disparate agents involved. There is a special term in MolochDAO, with corresponding software code, for when someone leaves in frustration: "ragequit," a term derived from gaming culture. Unlike the usual charity or early-stage investment, contributors can withdraw their stake at any time. Other DAOs have since adopted the feature. The platform DAOHaus enables users to easily start new entities on the MolochDAO template, complete with the same vernacular of summoning and tributes.

The working theory underlying this experimentation has come to be known as *cryptoeconomics*—a synthesis of economic incentives with cryptographic technology.[48] The term is widely associated with Ethereum founder Vitalik Buterin, who has written about cryptoeconomics as a nearly universal engine for social and technical processes. As Buterin puts it, cryptoeconomics allows software "to reduce social trust assumptions by creating systems where we introduce explicit economic incentives for good behavior and economic penalties for bad behavior."[49] For example, Bitcoin's cryptographic math protects the scarcity of units on its ledger; the perceived value of those units, in turn, motivates users to expend computing energy to perform expensive cryptographic math problems to win rewards. The math secures the economy, which in turn motivates people to use the math.[50] With the advent of Ethereum and its programmable "smart contracts," such a carrot-and-stick design extends from not just the management of an asset but also to the governance of countless applications, from financial contracts and art markets to social-media networks and philanthropic ventures.

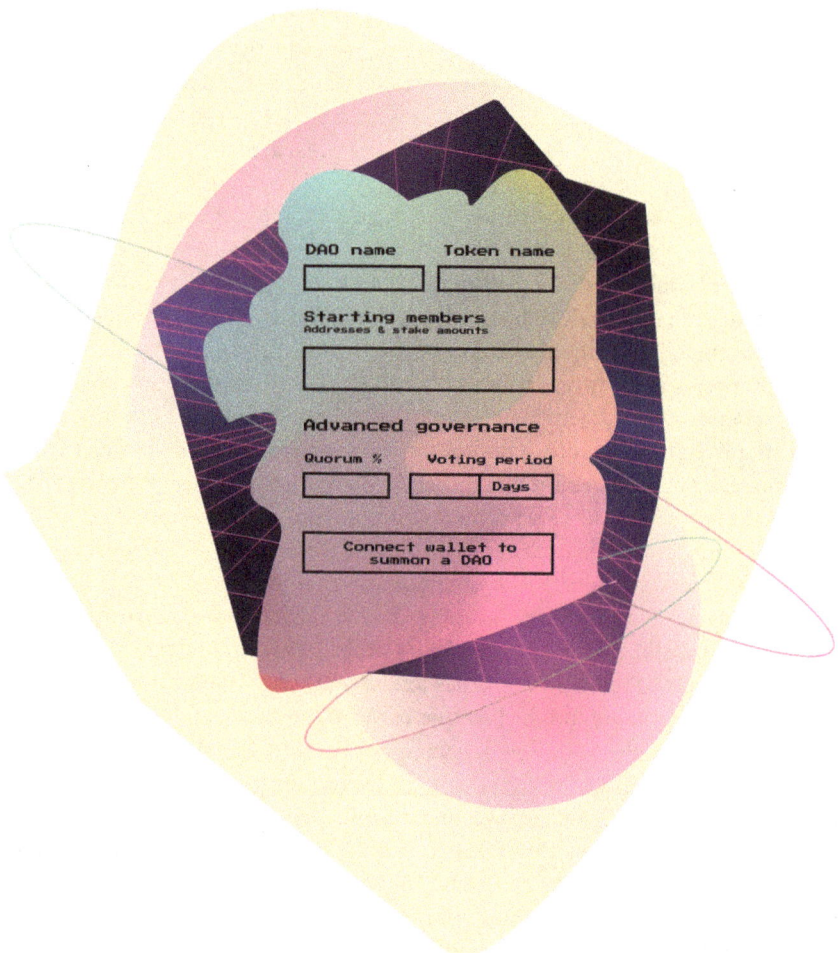

DAO name   Token name

Starting members
Addresses & stake amounts

Advanced governance

Quorum %   Voting period

Days

Connect wallet to
summon a DAO

FIGURE 9.

Participants differ over the meanings they ascribe to their cryptoeconomic media. A nonscientific survey of political views in crypto[51] identified not just "leftist" and "libertarian" tendencies but also positions that have nowhere to live on conventional political spectrums, such as the "Zamfirist" and "Walchian"; each is a matter of opinion specific to how power should flow on and around blockchains. They are high-stakes positions, as billions of dollars' worth of tokens may hang in the balance. Although crypto has drawn on certain earlier political lineages— particularly anarcho-capitalism and libertarianism[52]—as time passes, its ideological space becomes less reducible to those and more fecund for its own breeds of politics. The politics seem to evolve quickly in a world whose memetic repertoire spans transcendent registers from the Genesis Block to the hoped-for slaying of Moloch.

Crypto enthusiasts at times describe themselves as LARPing, or live action role-playing, like the hobbyists who dress in medieval costumes and hit each other with foam swords in a park. There are many promises among crypto's true believers that the technology's actual usage has fallen well short of: banking the unbanked, undermining financial elites, and defeating authoritarian censorship, for example. The technology doesn't exactly work as intended yet and maybe never will. Still, it is a medium and a gateway. Inhabiting a distinctive mythological, symbolic universe provides a license to dispense with foregoing constraints, which arose for a world connected with different sorts of media. The mythology then permits a transition from economics to cryptoeconomics, from politics to coordination, from representation to decentralization.[53] Moloch and his ilk are Tocquevillian gestures toward a provisional and only half-serious heaven, one bearing transcendent commitments that hold the early adopters together while their new technology aids in reshuffling the institutional tables here on earth.

Democracy doesn't feature prominently in the idiom of many crypto subcultures, perhaps out of fear that it will shake away the necessary reverie and bring everyone back to the still-looming regimes of territorial governments. Lately, *coordination* is the nearest surrogate, and it risks discarding democracy's preference for the common good against plutocracy. Yet the Moloch slayers do not only think and write about alternative voting systems, as Lani Guinier did; they test them out with friends, strangers, and the digital equivalent of millions of dollars—then notice the outcomes, the disasters and happy accidents, and fork the code to try again. The technology is dangerous; it can enable unaccountable pump-and-dump schemes as much as collective ownership. The specter of Moloch serves to spur the machine-makers to focus on the side of light, of coordination, of the common good. But it seems to me that the human-eating Moloch could just as easily reside in these new systems designed to vanquish him as in the principalities and powers that still rule the world.

Cryptoeconomics as a Rupture and Limitation

Regardless of any practical use value, crypto represents a rupture with respect to the particular argument I have been making: it can be an antidote to implicit feudalism. Previous internet technologies have presumed a central server, whose legal owner holds ultimate responsibility for what takes place on that server. A democracy among users will almost inevitably come into conflict with the underlying technical and legal reality. The distinguishing affordance of a blockchain, however, is enabling a system that lacks any single owner, that is user-governed by default.[54] Blockchain protocols differ from earlier networking protocols in that governance is embedded; the protocol defines how to change the protocol. The Bitcoin blockchain is thus designed to be governed by the users who secure it by "mining," while many other blockchains give power to those who "stake" tokens. DAOs typically confer governance rights on their token-holders. At the technical level, these systems are designed precisely to avoid the concentrations of feudal power that centralized servers have encouraged.

The technology's design, however, does not guarantee its social outcomes. Concentrations of power have been chronic in the governance of blockchains and organizations built on them. Crypto has been a tool for aspiring authoritarians, like El Salvador's Bitcoin-enthused president Nayib Bukele. Yet for reversing implicit feudalism, specifically, the opportunity is real.

Accordingly, crypto has occasioned a remarkable outpouring of democratic mediums that implicit feudalism fended off in the past. Tucked among the speculative bubbles, scams, and rampant financialization of crypto are countless experiments in online governance, more than in any previous period of internet history. The high financial stakes and the lack of external regulation surrounding blockchains have brought investments of tokens and time into the search for good governance. Because blockchain activity is by its nature public, the code behind successful experiments quickly spreads as others copy and adapt it. As a result, crypto has occasioned governance innovations such as

- decision-making processes that evaluate preferences in nearly real time,
- voting systems unavailable in conventional politics or business,
- mechanisms for incentive alignment among diverse participants,
- algorithmic dispute resolution,
- permissionless participation,
- widely shared accountability and distribution of benefits,
- self-enforcing security and censorship resistance,
- sovereignty from external control or regulation,
- transparency of on-chain activity,
- competitive markets for governance,
- ease of exit and capacity to fork systems,
- identity systems under user control, and
- novel interfaces for governance activity.[55]

In each of those are many particular examples. Among the decision-making processes, for instance, are these:

- *Conviction voting*: Votes on a proposal are continuously weighed based on both quantity of tokens staked and duration of staking.
- *Curation market*: Curators are rewarded for elevating proposals or projects that they correctly predict others in the community will like.
- *Decentralized dispute resolution*: A random jury of users with staked tokens independently choose the outcome of a dispute that they expect most others will pick, gaining or losing tokens based on their choice.
- *Lazy consensus*: Users with sufficient reputation from past activity can make proposals that pass automatically in the absence of objections.

- *Liquid democracy*: Token-holders can delegate their voting power to other trusted holders who can delegate them in turn; delegation can be withdrawn at any time.
- *Quadratic funding*: Matching grants are distributed according to a combination of the number of donors and the amounts they give to a cause.
- *SplitDAO*: A subset of users in a DAO can withdraw their tokens and move them to a duplicate entity.

The proliferation of governance techniques is not an end in itself, of course. It does not guarantee that governance will be in any way good. But, to borrow a biological metaphor, variation is a prerequisite for natural selection. For communities to identify the governance practices that work for them, they should have enough range of motion to explore diverse options. This range of motion is precisely what implicit feudalism has restricted in online spaces and what established forms of liberal democracy have often restricted in territorial governance.

Crypto has made unusually explicit what has always been true: money and other forms of capital are themselves media—malleable and programmable transmitters of information, which obtain value through the meaning-making they enable.[56] Blockchains, and the cryptoeconomics in their designs, are infrastructures for economic media grounded more in networks than in state power. Self-enforcing software can operate through economic stake and incentives in place of the monopoly on violence that state-backed financial systems employ. In a 2018 sci-fi-drenched promotional video for Aragon, a platform for blockchain governance, co-founder Luis Cuende boasted, "Today, we are in the first time in history that we can actually try out new governance models without the need of people getting killed."[57] While this claim is less true than Cuende thinks, there are respects in which the explorations at hand are distinctly novel.

If cryptoeconomics is the sole basis of new governance models, however, there is cause for worry.[58] Diverse voices have long warned against the expansion of economic logics, crowding out space for democratic politics in public life. From the Zapatista insurgents of southern Mexico to political theorists like William Davies and Wendy Brown, the neoliberal aspiration for economics to guide all aspects of society represents a threat to democracy and human personhood. According to Brown, "as an economic framing and economic ends replace political ones, a range of concerns become subsumed to the project of capital enhancement, recede altogether, or are radically transformed as they are 'economized.' These include justice (and its subelements, such as liberty, equality, fairness), individual and popular sovereignty, and the rule of law. They also include the knowledge and the cultural orientation relevant to even the most modest practices of democratic citizenship."[59]

The things not visible to the market, that is, become unthinkable. The market dictates a neoliberal people's range of options. If the market cannot see a changing climate, there is no motivator for acting on it. If the market does not recoil at the plight of homelessness, neither can we, if we learn to be what the market sees in us.

Worries about the corrosive possibilities of economics on politics preceded the terminology of neoliberalism. Hannah Arendt observed that ancient Greek democratic thought regarded economics as housekeeping, a private matter segregated from the political sphere.[60] Athens' sexist, slaver economy enabled citizens to enter politics as relative equals, whose "prepolitical" basic needs were already met, whose democracy could stand aloof from self-interest and corruption. To be a free and trustworthy citizen meant being free from susceptibility to economics, someone trustworthy when contemplating public life. In the realm of the political, for Arendt, people become capable of acting in truly new, truly creative ways.

Arendt's account of a politics wholly distinct from economic necessity provides a useful foil for our purposes here. Let my use of *politics* refer to some approximation of Arendt's: public action concerned with the common good. Political institutions are domains for Homo sapiens before Homo economicus—or, perhaps more relevantly, the Homo speculans of speculative finance. Even Arendt would not remove politics fully from economic life, since politics should shape the economic order, and it depends on that order. But what distinguishes politics is its capacity to notice and address considerations beyond the allocation of resources and to organize economies accordingly.[61] While a country's taxation policy utilizes economic nudges, for instance, lawmakers must generally rationalize it according to conceptions of the common good, rather than solely optimizing for financial metrics. Politics is hardly immune to self-interest.[62] But incentives such as the need for politicians to run for reelection can introduce imperatives that economics alone would not.

The limits of cryptoeconomic design and the need for political spaces have become increasingly apparent in actually existing crypto. If the purpose of a governance system is to enable participants to have as much self-determination as possible—a tolerable oversimplification, I hope—whatever inhibits that self-determination becomes a limitation. As crypto matures, its designers have learned that older concerns about the corrosive effects of economics on democratic governance are also relevant to distributed ledgers: rampant plutocracy, the suppression of participant interests, and dangerous externalities.

The first limitation, plutocracy, is a direct outgrowth of governance that arises not from personhood but from economic stake—whether it be through token holding or "mining" with expensive computational power. The power of concentrated wealth over human participants has been a growing anxiety in crypto networks, including among leading developers.[63] Of course, governance by economics is nothing new; joint-stock companies conventionally operate on plutocratic governance—more shares equal more votes. Yet companies exist within the constraints of state policy, which can impose counter-pressure like progressive taxation, collective bargaining rights, environmental regulations, antitrust enforcement, and more. If distributed ledgers are based on cryptoeconomics without an underlying political order, such options are not available. As long as governance is reducible to economics, it will be difficult to prevent the feedback loops between wealth and power from spiraling into plutocratic outcomes.

A second limitation occurs in what is thinkable and speakable among human users. Like economics itself, cryptoeconomics is normative as well as descriptive. People begin to cast themselves in the image of the systems they inhabit. According to one study of management education, a field that tends to regard human nature as competitive and acquisitive, "self-interested behavior is learned behavior, and people learn it by studying economics and business."[64] Systems are all the more constraining when they involve highly structured algorithmic processes, as crypto protocols generally do. While introducing algorithms may add efficiency to governance, a recent analysis finds that doing so can also result in "decreasing the space for governing actors' discretion."[65] It is no surprise, then, that the cultures surrounding crypto are highly attuned to algorithmically mediated economic indicators—using references to "bull" or "bear" markets to describe people's emotional states. But human beings have interests not reducible to economics, never quite encodable in an algorithm. Reliance on cryptoeconomic governance risks losing sight of other things important to human flourishing.

The final limitation appears in crypto's externalities, its effects that are invisible to its own internal processes. Bitcoin is governed most of all by the "miners" who carry out its computation—consuming energy at the scale of a mid-sized industrialized country. Miners often stand to benefit from ignoring their carbon footprint. A busier network roughly correlates to higher energy consumption and a higher trading price, increasing the value of the miners' rewards. Other externalities relevant to blockchains include money laundering, dealings in dangerous drugs and weaponry, tax evasion, and ransomware attacks on public infrastructure. If crypto's importance continues to expand, so does the danger of its potential for facilitating harm against people and the planet. The previous generation of Web technology has facilitated massacres and election interference; a new Web of money and contracts could get more dystopian quickly.[66] There must be safeguards that can counteract economic self-interest.

Another example of an externality is "public goods," or the shared, critical infrastructure that many participants in a system rely on but few profit from, the virtual equivalents of roads and bridges. Funding these essentials has been a persistent challenge for crypto builders, as they discover firsthand that market mechanisms alone fall short.[67] Before cryptoeconomics, non-market institutions such as governments and (at vastly smaller scales) charities have been necessary; crypto projects are reinventing them through fee-funded treasuries and donor grant pools. MolochDAO was a step in this direction, and other experiments have followed. But project after project continues to find that economic incentives alone are inadequate to generate healthy markets, to say nothing of goods that markets cannot provide.

The limitations I identify are exacerbated by the difficulty cryptoeconomics has in recognizing human identity.[68] This is a persistent but not necessarily permanent condition. Cryptography obscures users from each other; economic designs care less about who users are than the tokens they hold. Those tokens are how cryptoeconomics enforces its rules. Personhood is not a built-in concept for blockchains, as it is for any government with citizens. But the premise of any democratic structuring probably needs to be some way for systems to identify and represent individual human beings, along with relevant nonhuman agents.

While it has enabled productive experimentation, cryptoeconomics cannot serve as a sufficient basis for the governance possibilities in online spaces. My argument is perhaps anticlimactic in comparison to a technology that inspires such radical aspirations for remaking the world: crypto needs to rediscover politics. This entails enveloping economics within rules set by institutions not primarily economic in nature, which are capable of articulating, instantiating, and evolving shared understandings of the common good.

Already, crypto-governance practice appears to be reinventing some old wheels of institutional life, including the rudiments of politics. There are juries forming to resolve disputes, covenants enshrining shared values, and voting systems designed to reflect not just wealth but degrees of preference and the breadth of popular support.[69] Economics remains central to these; the systems typically enforce good behavior by requiring participants to stake tokens that they stand to lose. Yet the appetite is growing for crypto to recognize, when appropriate, the identities of the actual human beings who use it. Leading among those voicing this appetite is the apostle of cryptoeconomics himself, Vitalik Buterin. He has called for a new design paradigm and mythology, borrowed from the nomenclature of *World of Warcraft*, his once-favorite online game: *soulbound*.[70] Soulbound tokens, for instance, are not exchangeable commodities but remain with a particular user. Blockchains, it seems, need to account for souls.

Pairing cryptoeconomics with intentional politics can help overcome the limitations that bedevil cryptoeconomic governance alone. This does not mean that political mechanisms must occur in every app and protocol. Even standard liberal-democratic theory permits diverse forms of association and business within a democratic structure, and similarly politics may be necessary only at key leverage points in a crypto network. Economics has its place, but citizens make the market's rules through their civil rights rather than their economic power. Similarly, if democratic structuring were present at the base layer of crypto systems, participants could assert interests and externalities that cryptoeconomics alone would tend to obscure.

Some have argued that the proper means of democratic structuring for crypto is through state regulation.[71] This is happening to the extent that governments are intervening with taxation, securities enforcement, and even the development of their own digital currencies. But relying solely on existing governments inhibits at least some of what cryptoeconomics promises, such as the ability to experiment with radically diverse organizations, permissionless participation, and censorship resistance. Dependence on territorial regimes also lessens the capacity of these technologies to enable equality among users across borders. The actual democracy present in governments may be limited or nonexistent. Crypto can be a medium for political imagination of a kind that governments are unable to provide on their own, whether due to outright hostility or mere intransigence. It is probably neither sufficient nor desirable to outsource crypto's politics to governments.

An alternative is to incorporate democratic design into blockchain protocols themselves, or the apps and DAOs built on them. For this, there is much to learn from the legacy of cooperative business, which blends person-centric governance (one member, one vote) with market-based incentives (patronage dividends in proportion to participation). The cooperative model has historically enabled activities that many DAOs pursue, such as gathering capital from participants who are not wealthy investors, enabling them to hold meaningful governance rights, and distributing rewards fairly.[72] Cooperativism also provides a framework for democratic governance that can help counteract plutocratic tendencies, while putting noneconomic values at the center of decision-making. Recognizing this, a growing number of DAOs have been incorporating as cooperative legal entities. Others are integrating co-op values into their software, such as by reserving governance power for workers or active users.[73]

A further strategy for democratic structuring is to establish a robust set of rights, responsibilities, and guarantees. Crypto networks have already developed constitutional layers in the code of their underlying protocols, which protects certain rights such as property and censorship-resistance.[74] A much wider set of values could be encoded into protocols, such as ones that prevent harm to people and the natural world, enforceable through the protocols themselves. Natural language agreements can be enforced through cryptoeconomic courts.[75] Future protocols might include code that ensures certain protections for workers, prevents direct harm to humans, or guarantees a basic income to all users. Protocols might ban carbon-emitting miners and other ecological harms. Rights-based designs could counteract plutocracy and make externalities more visible to a protocol that would otherwise ignore them. Cryptoeconomic designs can thus achieve goals not reducible to maximizing wealth.

The actual track record of activity on crypto networks has proven unnerving. Viral tokens attract hopeful retail investors, who may reap astonishing gains or see large chunks of their savings wiped away in hours. Venture capitalists meanwhile hold large stakes in important DAOs and protocols through which they can dominate governance processes. Perhaps, as Lana Swartz suggests, the whole point of crypto is not the promise but the scam, the "arbitrage on uneven belief among participants" in a hoped-for future "ever coming to pass."[76] Any opportunities that crypto presents on behalf of democracy, therefore, accompany opportunities for democracy's enemies.

The billion-dollar stakes of the crypto ecosystem have been uniquely generative for stress-testing experiments. Buterin writes, "Crypto is the ultimate training zone: if you can build something that can survive in this environment at scale, it can probably also survive in the bigger world as well."[77] As this training zone develops, participants should be attentive to how their systems succeed not only in enabling functioning markets but also in achieving at least as much justice as the systems they are supposed to replace.

DEMOCRATIC PLAY

I once attended a small party of Moloch slayers at a junkyard-themed venue in Colorado. Buterin was there, along with leading Ethereum entrepreneurs and a documentary director with a camera recording on and off. Conversation veered between stilted small talk and ferocious discussions on the merits of various blockchains. The evening didn't really seem to flow until, several hours in, the chess boards came out, side by side in a row, each with a clock next to it. I am not much of a player, so I watched as most others present took their places, and the room discovered an earlier elusive clarity of purpose.[78] It occurred to me that maybe the whole undertaking of crypto, for its architects, served the same function as a game of chess: an absorber of mental computational cycles for people otherwise bored by the normal world and a test of prowess among them. The blockchains beginning infiltrate high finance and pop culture, perhaps, were for the people building them only the latest fascinating game-board.

Then it would be no accident that Buterin has explained his affection for decentralized protocols with a story of teenage agony—when the centralized corporate owner of *World of Warcraft* issued a unilateral software update that messed up his progress.[79] Crypto communities frequently form on Discord, a social platform first developed for online gamers. The beating heart of a crypto Discord server is the meme channel, where members try to one-up each other with jokes and playful propaganda, only some of which make their way to the more public X and Reddit feeds. The quest to defeat Moloch, which I have so far treated as a kind of grand mythology, is also just another meme game.

Crypto would not be the only sort of high stakes politics to seem, if you squint your eyes just right, reducible to play. From the ball courts in ancient Mayan temple complexes to the partisan contests among political operatives in present-day elections, the playing of games is never far from the machinations of power. It happens that my first encounter with the classic treatise on play and culture, Johan Huizinga's *Homo Ludens*, was as pre-reading for a conversation on governance among crypto enthusiasts.[80] *Homo Ludens* is a mighty feat of World War II–era European erudition, traversing a wide sweep of ancient history and colonial ethnography. Huizinga concludes that play, within rules set apart from other realms of life, is the engine from with culture arises and on which civilization depends. He traces the rituals of Indigenous societies that respond to harm without outright punishment, using dance, dress, and choreographed jousting to resolve what might otherwise be persistent cycles of revenge. Vestiges of such play persist in the scripted, costumed rituals of modern courts, although Huizinga finds that coercive power and punishment have there supplanted most of the capacity for playful conflict resolution.[81] As the war against Nazism raged around him, Huizinga identified the fascists' mass rallies and goose-stepping as a sort of false play, not exploring human possibility but suppressing it.

Decades later, in the context of transformative justice, adrienne maree brown takes play a step further with the language of *pleasure*. She reminds fellow activists that struggle cannot be all serious and dutiful, that social change should revolve around what makes life nourishing, delicious, and erotic. "We all need and deserve pleasure," she writes, and "our social structures must reflect this."[82] Part of how oppression works, she observes, is by granting some people the privilege of enjoyment and denying it to others, whether through outright barriers to spaces of play or through impositions of shame.

The cases in this chapter involve dramatic differences in who has the privilege to play. Through speculative investment in their still-experimental technologies, crypto builders have been able to play with financial resources unavailable to activists in over-policed, under-resourced neighborhoods. Some sites of imagination have far more to work with than others. In a just society, all people should have pathways to play and the political possibilities that come with it. Demanding and protecting the right to play is a struggle to more fully self-govern.

Here, brown's fractals are at work again—the interweaving of everyday life and political power. Transformative justice has never been about just municipal police budgets; it begins and ends with people experiencing safety differently in their communities. Crypto, meanwhile, cannot be understood as merely a new class of digital assets; its value has always been bound up in the making of new mythologies and remixing old ones. In both imaginative worlds, play is a bridge between present institutions and future ones, between Tocqueville's "earth and heaven." Through play, people cultivate the political skills that the next evolution of democracy will require.

The political economist Vincent Ostrom, a twentieth-century disciple of Tocqueville, taught that citizenship is a sort of "artisanship," both art and science. Unlike a sculpted pot or statue, however, Ostrom noted that organizations are "artifacts that contain their own artisans."[83] The possibilities of organizational life are not to be found in universal laws but in the political imagination their participants allow themselves to have, and in the capacity to sneak imagination into practice.

This chapter has begun to chart a departure from the logics of implicit feudalism and homesteading with two very different case studies in mediated political imagination. Transformative justice and cryptoeconomics both challenge existing institutions through transcendent reference and immanent play. In so doing, participants have been able to explore democratic mediums still foreign to the reigning social infrastructure. From here, I turn to the preconditions and substrates that could make such practices much more widespread. How can communities gain the power they need over their online spaces to engage in democratic play?

The Projects

SHORT READ   DEEP READ

EXCAVATIONS
GOVERNANCE ARCHAEOLOGY
FOR THE FUTURE OF THE INTERNET

The Taxonomy   Consent   Assembly   Mandate

FIGURE 10.

Excavations

excavations.digital

Excavations: Governance Archaeology for the Future of the Internet is an online exhibition that presents work from a residency involving ten artists and collectives. The residency took place online during the height of the COVID-19 lockdowns, with artists located in many parts of the world. During the residency, they reflected together on the present and future of online governance, grounded in discussions about the long history of human societies. At the end, we exhibited *Excavations* at the United Nations Internet Governance Forum.

The residency coincided with the development of a database that collects historical, cross-cultural, collective governance practices, discussed in chapter 4. Conversations with the artists informed the design and interpretation of the database. In *Excavations*, the artworks appear with links to patterns in the database's preliminary taxonomy.

The New Delhi–based collective Barabar, for instance, developed a work of dystopian fiction called *The Rights Market* that imagines human rights available for purchase à la carte on a convenient app—echoing patterns in the database of "auction" and "consent." Mateus Guzzo produced a diagram called *Public Audio*, a representation of the Brazilian media ecosystem inspired by Salvador Allende's Project Cybersyn in 1970s Chile; it evoked historical patterns of "monitoring" and "positive reinforcement." Haudenosaunee artist Amelia Winger-Bearskin worked on *SKY WORLD/CLOUD WORLD*, connecting cloud-based chatbot technology with Indigenous conceptions of honoring the sky, reflecting patterns such as "matriarchy" and "reparations."

Many of the *Excavations* artists came to their work with a strong sense of ancestry—of obligation to ancestral lineages, both direct and adopted. This sensibility has helped inform our thinking surrounding the meaning and use of the database. If the database were simply another act of appropriation, the artists taught us, it should not exist at all. The information it contains must, rather, be an invitation and starting point for relationships, for accountability, for repair.

4

Governable Stacks

Organizing against Digital Colonialism

While the island was still a French colony, in 1801, Saint-Domingue's slave revolter turned governor-general Toussaint Louverture convened a national assembly. Later that year, it proclaimed—and at substantial cost, printed—a constitution describing a nominal French territory ruled by the former slave's government. Writes C. L. R. James, "To have it printed meant (in those days) that an irrevocable decision had been taken."[1] Louverture's brazenness made him intolerable to the colonizers. The following year, Napoleon's troops deported Louverture to France, where he died in prison as the French waged a doomed, vicious war to regain the island. On the first day of 1804, Louverture's successor Jean-Jacques Dessalines founded the independent state of Haiti.

The film *Finally Got the News* depicts the League of Revolutionary Black Workers in Detroit, a left flank to the United Auto Workers that identified with the liberation movements spreading across Africa since World War II.[2] The League was a militant organization, enmeshed in the city's violent uprisings during that period. But in the film what we see is not burning city blocks. An organizer speaks from behind a desk in an office, surrounded by what one imagines to be membership rolls and correspondence in progress; members hand out leaflets to fellow workers at the door to their plant.

In both scenes of liberation movements, self-governing coincides with intentional media use. This chapter considers the governance of online space as another site of resistance against domination. Creating spaces governable by their participants is not simply a matter of exiting to a new homestead on some endless digital frontier. That frontier and its homesteads were fictions all along, while platform companies gained growing control over the finite time, space, cultures, and economies of the embodied world. As in Haiti and Detroit, self-governance requires communities to take control of the technologies with which they organize.

Critics have been converging around the language of colonialism to describe the internet economy, using no shortage of terms: digital colonialism, technocolonialism, data colonialism, data orientalism, digital capitalism, digital extractivism, platform imperialism, postcolonial computing, decolonial computing, and imperial play, for example.[3] Computing, writes Syed Mustafa Ali, "is colonial through and through." Stefano Harney and Fred Moten identify a lineage from the Atlantic slave trade to the packet-switching of ARPANET: "the dream of this newly dominant capitalist science" in which containerized logistics packages every part of life into the possibility of being "shipped."[4] Less developed than the critiques, however, are the means of resistance.

I will use *digital colonialism* as a capacious shorthand for the above terms— forms of domination by governments and corporations through their control over internet technologies. I do so while recognizing the danger of too easily conflating military occupation with more immaterial feats of data extraction and digital labor arbitrage. In the apt phrase of Eve Tuck and K. Wayne Yang, "decolonization is not a metaphor."[5] Corporate capture of online data is not the same as territorial conquest and genocide. But the control of data flows can supplant or aid control over embodied life. To the extent that access to livelihoods and cultural sovereignty occur through digital systems, the coloniality in question is no mere metaphor. Online life, too, is a site of struggle. And if we are serious about the laden language of the colonial, we should be ready to learn from past struggles against pre-digital colonial regimes.

Alongside acts of outright insurrection, theorists and practitioners of anticolonial resistance have articulated the centrality of self-governance in everyday life for their movements. Meanwhile, the aspiration to be "ungovernable" has appeared among thinkers ranging from European philosophers Michel Foucault and Giorgio Agamben to former Black Panther Lorenzo Kom'boa Ervin, each seeking to assert the vital personhood of people caught in dehumanizing systems. Such systems of "governmentality" extend their power into subjects' lives through daily life, imposing order through habits of practice and thought.[6] Yet, I will argue, anticolonial traditions teach that ungovernability alone is insufficient as the basis of either resistance or liberation. It must accompany what the Honduran Indigenous activist Berta Cáceres called "decisive democracy": communities with the means to determine their own futures.[7] To become ungovernable under digital colonialism, in particular, how should we be learning to self-govern?

I pose this question in light of implicit feudalism. Tools for basic group decision-making are not widespread, nor are mechanisms to hold those in authority accountable. The design of social platforms inclines toward enabling the governmentality of platform owners, aided by their user-administrator proxies, rather than user governance that could turn against the owners' interests. Campaigns of digital resistance often employ the same colonial platforms whose hegemony they oppose.

Both settler colonialism and digital "user experience" involve regimes that dictate who has the right to self-organize, or not, and under what conditions.[8] Micro-targeted discrimination singles out individuals for exposure to exploitative product ads. The same targeting also inhibits public outcry. Algorithmic decisions about welfare checks and prison sentences make it harder for harmed communities to put collective pressure on individual decision-makers. Humanitarian organizations collect data about refugees, which the refugees themselves cannot access, while the organizations use it for future fundraising. Individual users of a platform might be able to see or delete their personal data, yet platform companies alone can analyze and monetize the data of the communities they host. Platforms impose the developers' cultural norms, projecting a false universality that leaves little space for user communities to practice their own cultures. And at least as much as platforms might enable activist organizing, they introduce new varieties of surveillance and repression.[9] People confronting digital colonialism today might resist these kinds of incursions, following past anticolonial struggles, by rediscovering and reinventing the art of self-governance.

This chapter contributes to the design of networks that refuse colonization through self-governance. As a bridge between struggle and fulfillment, I introduce the concept of *governable stacks*: the interconnected infrastructures and practices that enable networked self-governance. Next, a design paradigm of *modular politics* outlines how governable stacks could replace implicit feudalism. I then turn to *governance archaeology*, the work of filling governable stacks with lessons from ancestors across diverse times and places.

"Governance is what we are fighting for," writes Black Lives Matter co-founder Alicia Garza. "We are fighting for the right to make decisions for our own lives and to ensure that right for others."[10] This is both the goal and method for movements around the world, often facing daunting odds. But self-governance is no guarantee of more just outcomes; authoritarians are building stacks of their own, which they can govern as they see fit. The governable stacks that people craft intentionally today can be the basis of a future where democratic online spaces are everywhere we need them.

"TO STRUGGLE AGAINST GOVERNANCE"

Governance talk does not always sit easily with movements for liberation. "Governance is the extension of whiteness on a global scale," write Stefano Harney and Fred Moten.[11] NGOs are the "laboratories" of governance, which use the rhetoric of democracy to uphold order through the guise of humanitarianism. This governance is a cheap sort of domination because the subjects do it to themselves: "Governance arrives to manage self-management, not from above, but from below." Harney and Moten call instead for a politics of refusal and "being without interests," a call to imagine what it would mean "to struggle against governance": "We are the general antagonism to politics looming outside every attempt to politicize, every imposition of self-governance."

Harney and Moten can claim many precursors. They frequently invoke Frantz Fanon, who admired the "spontaneity" in popular uprisings, the ungovernable reaction of the lumpenproletariat, "the most spontaneous and the most radically revolutionary forces of a colonized people."[12] They evoke the ungovernable villages of escaped slaves in the Americas, including the maroons of Saint-Domingue's high hills, whose raids did not wait for Toussaint Louverture's command but made possible the eventual independence of Haiti.[13]

"You know, I love C. L. R. James," says Moten in passing.[14] James, the Trinidadian chronicler of Louverture's revolution and an instigator of others from Tanzania to Detroit, praised spontaneity as well. His 1958 book with Grace Lee Boggs and Pierre Chaulieu, *Facing Reality*, describes a "most conscious and finished opposition to the parliamentary procedure" found among dockworkers. By their account, "dockers do not like votes"; "they sense the general sentiment and act on that."[15] What holds sway is a worker's je ne sais quoi ability to capture the attention of the others, regardless of role or position: decision without institution.

The age of networks has only deepened the allure of spontaneity among radical theorists, as in Michael Hardt and Antonio Negri's celebration of the "multitude" and the "assembly" against fixed organizational forms or Manuel Castells's "networks of outrage and hope."[16] Underground tracts from such pseudonymous formations as the Invisible Committee and the Vitalist International long for rebellions whose disorder is their vindication, while adrienne maree brown, in the lineage of C. L. R. James and Grace Lee Boggs, presents spontaneous self-organization in nature as a theory of social change.[17] These thinkers seem to hold that the organizational forms of past revolutions no longer compute—especially because we now have computers.

An antithesis: Over a century ago, Vladimir Lenin regarded revolutionaries "who kneel in prayer to spontaneity" as a "fungus"—and not with any of the admiration brown would later hold for fungi.[18] Where there is spontaneity among the masses, it obtains power only through an organized and disciplined vanguard party, such as the one he would lead in Russia. Rosa Luxemburg recoiled at the rigidity of Lenin's vanguard, one molded by the discipline of the factory, the army, and the bureaucracy. She called for a movement that would be "supple as well as firm," capacious enough to hold the full humanity of its participants.[19] A communist regime came to pass in Germany, however, not through her homegrown movement but through Soviet tanks rolling into Berlin. Those tanks emanated from Stalin's dictatorship, as evidence that Luxemburg was right to worry about a vanguard modeled in industrial discipline. Yet what she longed for remains so often elusive: a movement firm enough to gain power while supple enough to wield it humanely.

Now stop and go back, and reconsider those apparently kneeling before spontaneous resistance, against the strictures of governance. Synthesize the dialectic. Fanon also warned against the "cult" of spontaneity and stressed that the "enlightening of consciousness" necessary for liberation is "only possible within the framework of an organization, and inside the structure of a people."[20] He held that spontaneous energies must find institutional cohesion. C. L. R. James affirmed, in his final interview, "I believe you must have an organization," in something like the Leninist sense. He celebrated the Paris Commune as a forerunner of the Russian soviets, regarding that uprising as "first and foremost a democracy." In "Every Cook Can Govern," an essay that took its title from a phrase of Lenin's, James recommends to workers the ancient Athenian method of ruling by sortition, selecting authorities from the citizenry by lot.[21] Struggle requires organization, that is, but it must be creative and accountable, reaching into the lives of those who self-govern through it and also outward as a model to others. Accordingly the independence movements James helped to inspire sought not just nation-states but a new order of global governance.[22]

Grace Lee Boggs was long a fellow traveler with James in the factions and divisions of sectarian Marxism, a student and friend of Third World revolutionaries. Through organizing in Detroit with her husband Jimmy Boggs, she thought her way into a "politics of personal development" that rejected partisan orthodoxies in favor of a more iterative "dialectical humanism," in which political visions and the people who hold them evolve together through struggle. Later in life, she studied ecology and the dynamics of systems more complex than mere dialectics. As she drifted from Leninism, the centrality of self-governance only deepened. She became a mentor to veterans of the 2011 Occupy Wall Street protests, following their "leaderless" experiments in radical consensus. Her orientation turned from achieving state communism to commoning, the work of stewarding shared projects and resources in relationship with their natural environments.[23]

The influence of Boggs has continued to spread since her death in 2015, at one hundred years old. Political theorist Rodrigo Nunes has envisioned post-2011 movement organizations with Boggsian, naturalistic language like "nebula" and "ecology." He confesses attempting to recuperate a kind of vanguardism, a "networked Leninism"—before concluding with an insistence that above all, activists should "think and act ecologically."[24] In Boggs we see the origins of passages about mycelia and butterflies and trees that recur in the writing of adrienne maree brown. Brown's "emergent strategy" for activists revels not in conflict with corporate opponents but in apparitions of friendship in online threads and tips for weaving consensus processes. Seeking to transcend "protest politics," Boggs described her mentorship of younger organizers like brown as "projecting and initiating struggles that involve people at the grassroots in assuming the responsibility for creating new values, truths, infrastructures, and institutions that are necessary to build and govern a new society."[25]

Fred Moten acknowledges the Boggses' influence as an example of unpayable debts.[26] What he and Harney offer in place of governance is "study"—a term of art that is also resolutely plain, referring to the gathering and learning that takes place among groups of people in spaces ungovernable to reigning institutions. Like the maroons of Saint-Domingue or the American South, study surely involves an order of its own, apart from the colonial university, a practice of insurgent self-rule. The maroons of study, for Harney and Moten, are never-settled communities of exodus. But their maroons undertake "fugitive planning." They study to plan; they plan so that they can find the space and time to study. To do either and therefore both, there must be something of the self-governance Harney and Moten seem at first to disavow.

These legacies of resistance speak loudly the more you listen: to be ungovernable in any durable way requires self-governing through everyday organizing. Platforms have enabled their users to feel ungovernable and powerful for a time. But without the means of self-governance, those sensations will be always fleeting.

Virality as a Colonizing Strategy

I once entered the office of a labor organizer to find her with her head in her hands. She was running a campaign in the ever-shifting, just-on-time, atomized theater of urban retail. Why so down? The workers were migrating to Instagram. At least on Facebook, she could corral them into groups and post updates. On Instagram, every message had to be hilarious or enraging or gorgeous if she wanted it to reach them. Sometimes the information an organizer needs to share is not any of those.

Rather than persistent groups or organizational membership, Instagram's eminent form of shared experience is the viral image, which circulates an affective impression of shared experience. To spread, the image must be the kind of image that *would* spread, according to the tastes of the poster's followers and the secret churning of the platform's engineering. An announcement for next week's union meeting may not qualify. An organizer trying to strengthen workers' bonds isn't interested in infecting them like a virus.

The rise of ubiquitous social media rode on waves of protest. Individual voices, linked with hashtags, seemed to herald collective liberation. Protests spread on social networks like never before: the Zapatistas in 1994, the Battle of Seattle in 1999, Iran and the Tea Party in 2009, and then the wave of 2011 that began in the Middle East and spread to Europe and Wall Street. The Umbrella Revolution, Black Lives Matter, #MeToo, End SARS, Standing Rock, and so many others followed. Believing that the new social media rendered foregoing social structures obsolete, activists experimented with direct democracy at the scale of thousands. But after the exhilarating viral moments passed, the social media that radically democratic protesters relied on failed to support persistent organizations.[27]

Despite the outpourings of promise and hope and near-term victories, 2011's digitally mediated uprisings have fallen under the police of Mohamed Morsi and the bombs of Bashar al-Assad, the famines of the Yemeni civil war and the warlords of Libya. "Pirate" political parties arising out of online protest have tended to collapse upon their first encounter with power, if they ever got there.

At the Occupy Wall Street encampment, reporters would arrive and be transfixed by the media center—the nerve center, the center of power because it was the producer of media.[28] And media were powerful indeed, as they aided in drawing thousands upon thousands of people into what began as a small, precarious protest. Videos of police attacking activists bred sympathy and attracted participants, who began entertaining a feeling that the movement might be on the brink of sparking some kind of revolution. At least at first. By early the following year, the videos didn't work the same way. An activist monitoring the analytics data noticed at the time that "riot porn is losing its luster for mass online consumption."[29] As the social-media attention waned, so did the movement's influence.

A decade later, nearly all the viral movements of 2011 had succumbed to emboldened versions of the forces they had opposed. The likes of Abdel Fattah el-Sisi, Vladimir Putin, Donald Trump, and Xi Jinping discovered how to outlast digital insurgencies, obscuring outbreaks of dissent under a deluge of obfuscation. Virality is a commodity online, and armies can produce it for themselves. Zeynep Tufekci offers an illuminating distinction: the networked "signal" of movements can be self-defeating without "capacity" to translate it into durable, adaptable organizations that can wield leverage long enough to achieve shared goals.[30] For movements that claim a democratic mandate, capacity for power requires capacity for sustainable self-governance.

The classic strategy of colonial domination—*divide et impera*, divide and rule—proposes to dominate by training subjects to feel an illusion of power through their conflicts with one another. On colonial platforms, too, users joust for influence and affirmation, identifying themselves ever more deeply with the nontransferable reputation they obtain. Virality is fleeting if it ever happens, but the possibility is there, feeding what Jodi Dean has identified as a fetish of circulation, an end in itself that supplants goals for political change. Before long we have recapitulated the final scene of the 1954 McCarthyist blockbuster *On the Waterfront*, in which the dockworkers flee from their union's problems into the arms of the boss, newly able to experience their common exploitation as individual liberation.[31]

Virality seems to offer a sort of ungovernability in the relentless freedom to say anything and constitute momentary publics. But the economy of virality does not bow to the drudgery or necessity of self-organization. Platforms optimize for "engagement" through chatter—not decision, resolution, or consensus. Community control is not in the specifications unless communities put it there themselves.

SPINNING WHEELS AND GOVERNABLE STACKS

The actor Charlie Chaplin met Mohandas K. Gandhi in London in 1931. Chaplin later recalled that, after a bout of anxiety about what to say, he began, "I am somewhat confused by your abhorrence of machinery."[32] Gandhi explained that machines were not the enemy, the empire was. He spun his own cloth to resist the British textile monopoly in India, which controlled the processing of

Indian-grown cotton through English factories. The competition with industrial looms, backed by imperial decrees, decimated traditions of homespun textile production. (Europe's looms of the time were highly sophisticated technologies, containing in their designs critical precursors to digital computers.) Gandhi called for people across India to join him in spinning their own cloth on simple devices under their own control—an act of political, economic, and cultural self-rule. As he explained to Chaplin, Gandhi traded a machine out of his people's control for another they could use with dignity. Three years later, after hearing a story about factory conditions in Detroit, Chaplin had shed his earlier confusion and began work on the classic satire of mechanized capitalism, *Modern Times*.[33]

Today Gandhi holds a tenuous place in the anticolonial canon. His ever-evolving vision of national liberation fell short of liberation for all, particularly people facing subjugation by race, gender, and caste.[34] His demands on followers, beginning with his own family, could be ruthless and cruel. And yet Gandhi was an anticolonial leader who was both especially resolute in articulating a strategy of self-governance and successful in the work of dispatching foreign occupiers. His success inspired more struggles from Soweto to Alabama. And his teachings combined that confusing attitude toward machinery with the practice of creative self-governance.

The flag of the pre-independence Indian National Congress had at its center a spinning wheel, the symbol of Gandhi's "constructive programme": self-rule, or *swaraj*, as the basis of both resistance and the society that would follow. After independence, the flag lost the spinning wheel, but by law it still must be made of hand-spun cloth. Gandhi believed that self-sufficient and self-governing people would become ungovernable to colonizers. He regarded this, not the more famous and visible acts of protest, as the heart of his politics. "Civil Disobedience without the constructive programme," he wrote, "will be like a paralysed hand attempting to lift a spoon."[35] The link between self-governance and resistance was so strong for Gandhi that he regarded his personal self-control, even in diet and sexuality, as intertwined with the fate of the independence movement. He was interested in technologies that he saw as better suited for community governance.[36] The spinning wheel was a cipher with which Gandhi encoded self-governance into the Indian independence struggle—by his stubborn insistence on using a governable tool.

The spinning wheel remains a cipher, a site of conflict over the meaning of Indian democracy. Hindu nationalist prime minister Narendra Modi, despite having political ties with Gandhi's assassin, promotes homespun cloth; he has organized photo ops of himself operating a spinning wheel. Modi has meanwhile shuttered boards that gave actual artisans a voice in policy, under the slogan "Minimum Government and Maximum Governance."[37] The technology of the spinning wheel itself does not guarantee self-governance, but for Gandhi at least it was the symbolic base from which ever-enlarging acts of self-governance could defeat an empire.

In the spirit of this technological cipher, I propose the pursuit of governable stacks—the webs of tools and techniques that can support self-governing online communities. Governable stacks are cyborg assemblages of interoperating technology in symbiosis with human relationships.[38] Those relationships organize power in partnership with the technology more than through domination over it. Governable stacks are also an orientation toward ungovernable organizing under digital colonialism. They are the socio-technical substrate of governable spaces.

The geek-colloquial meaning of *stack* is a set of interoperating hardware and software. A tool higher up in the stack depends on those beneath it. Benjamin H. Bratton takes this usage further, describing the stack (or "The Stack") as "a new architecture for how we divide the world into sovereign spaces."[39] While he investigates The Stack primarily as medium of "planetary-scale computation," I want to turn our attention first to the stacks we experience at the scale of more immediate community. The planetary scale will emanate from those, but first of all a stack is a set of relationships. It might include all that enables one to use a social-media service, for instance: the server farms, the corporation that owns them, its investors, the software the servers run on, the secret algorithms that analyze one's data, the mobile device, its accelerometer sending biometric data to the server farms, the network provider, the backdoor access for law enforcement, and so on. The layers of a stack might further include the waterfalls or coal powering it, the wars fueled by rare-earth mining, and the mythologies and rituals that dictate what people in it will tolerate. Each layer is in fact multiple layers, and layers build on each other. The layers come with intersecting relations of dependency, along with emergent freedoms:

- *Community*: membership, codes of conduct, norms, rituals, relationships, economics, governance processes, histories, care work, education
- *Interface*: applications, servers, experience design, hardware, localization, usage constraints, access rules, operating systems, app stores, maintenance, repair, technical support
- *Infrastructure*: backbone networks, last-mile connectivity, government regulation, electricity access, network topology, legal ownership, corporate structure, hardware production, research and development
- *Ecology*: raw materials, health of workers and users, clean air, stable climate, resource-commons management

Recall how implicit feudalism spreads across the stacks where it occurs by filling power vacuums. Email is an open, decentralized protocol, but it has become dominated by a few companies who have used their friendly interfaces and market power to make the protocol a centralized dragnet. If a nondemocratic company holds legal liability at the legal layer of its stack, it will have to avoid running social-networking software that gives users enough decision-making power to conflict with its executives' control. The concept of the admin has spread from the design of server operating systems to the communities that arise on social applications. Centrally controlled technology has inspired a new breed of centrally controlled organizations. These layers of the stack could, in principle, operate in distinct ways; in practice they rarely do. Feudalism at one layer demands it of other layers. But if feudalism can spread across stacks, surely democratic designs could, too.

Before governable stacks were a concept, they were an experience for me, particularly through an organization in which I have been an anecdotal participant-observer for a decade. May First Movement Technology is a cooperative that provides Web hosting, cloud services, and public education for a 850-strong membership composed largely of activist organizations in the United States and Mexico. It is a descendant of the Indymedia movement, which pioneered social media practices in activist communities at the turn of the millennium. Through the tools May First offers, I have been able to move much of my daily computing away from companies that surveil and extract and into servers I co-govern, running freely available software. I have formed relationships with the people who maintain these services and participated in decision-making over bilingual conference calls and online ballots. I learn about new tools from fellow members, and we sponsor events that teach people outside our membership how to challenge the power of big tech in their lives and their communities. This is slow computing, its pace measured not by bandwidth or processing speed but by attention to the social dimensions of everyday practice.[40]

While Silicon Valley elites escape to phone-free retreats and agonize about their children's exposure to screens,[41] May First offers no such "abhorrence of machinery." It does not accept the false choice between addictive, surveillance-addled apps and a fantasy of returning to blissful innocence. Instead, members share technologies that do what they need and that they can reasonably control. These technologies and the self-governance we surround them with are our stack.

For me, being part of a governable stack like May First has unlocked political possibilities. The experience has motivated years of working to build governable stacks elsewhere, because I know that it can be done. With time, ungovernable stacks have come to feel like foreign lands. I often use them out of deference to other people's comfort zones, as well as to my employer's policies, but they never feel like home.

Technologists seeking alternative visions have often gravitated to the Free Software and Open Source movements, which employ creative licensing to enable the sharing of accessible and modifiable code. These movements have been successful in terms of the sheer volume of widely used software in their commons. May First relies on commons-based software exclusively. But the movements' emphasis on the freedoms of individual users, as well as of corporations, has privileged those with the technical know-how to take advantage. The software commons has spawned operating systems that fly in military jets and databases that aid in the imprisonment of asylum seekers. In the name of freedom, too, developers have harbored sexism and other forms of exclusionary culture.[42] Governable stacks should prioritize community accountability alongside individual freedom.

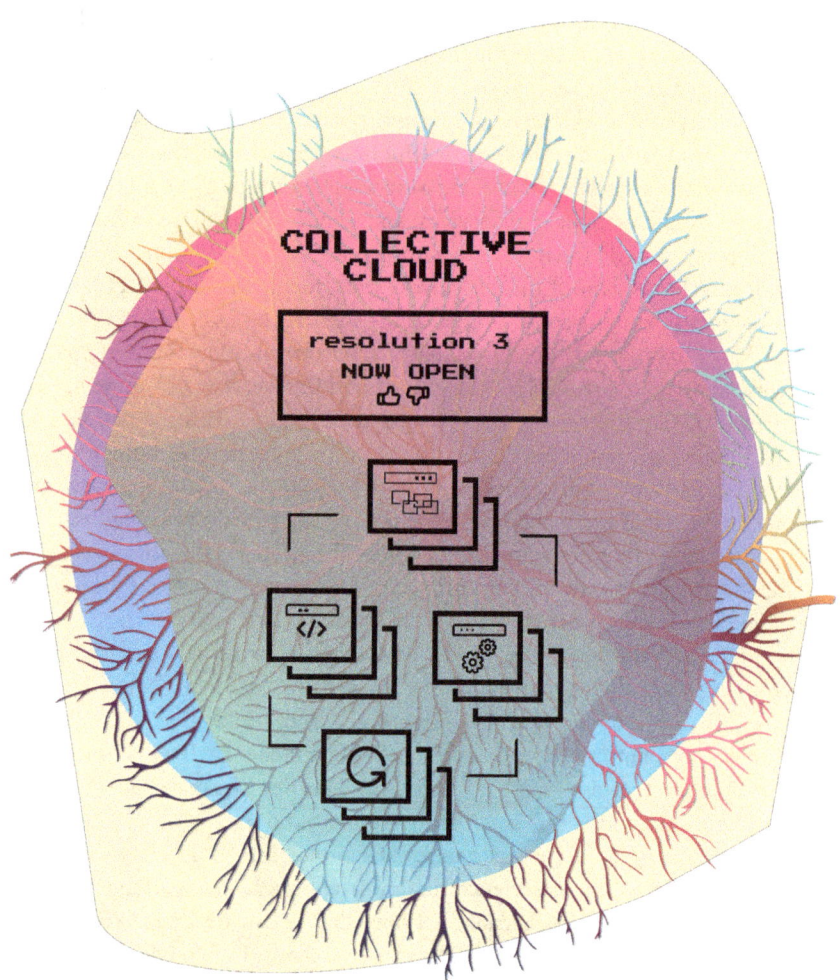

FIGURE 11.

Another strategy for challenging digital colonialism comes from labor power. Employees at Silicon Valley giants have achieved reforms by organizing against certain ethical outrages at their workplaces.[43] This can be a means of achieving greater governability for the communities those workers inhabit. Yet there are limits to what the campaigns are likely to achieve, since these workers are invested—often literally, through stock options—in the basic business models of their employers. Employees' actions can present the impression that their protest cleanses the colonial tools they produce. But governable stacks do not seek merely to improve the occupier. "Decolonization is not an 'and,'" as Tuck and Yang write. "It is an elsewhere."[44]

Experiences with governable stacks introduce us to possible elsewheres. The spinning wheel pointed toward an elsewhere—the invention of a democratic India—just as "feminist servers" in India today carry on and challenge that legacy, modeling a digital society free of patriarchy.[45] May First is an elsewhere for its members. Collectives, families, and movements can assemble and adjust their stacks over time, seeking to make their technological lives more governable wherever possible. Communities might go on using colonial platforms for education and organizing.[46] They might spread viral messages and enjoy what others share. But if they have governable stacks to go back to, they are more than just subjects. They are maroons, with swamps and forests of their own. There, they can imagine and work toward a world where they can be safe and powerful anywhere.

May First is infinitesimally small by the standards of the online economy. But spinning wheels are small, too, and they helped drive away the British Empire. Adrienne maree brown credits Grace Lee Boggs for helping her see the fractal nature of movements, that "what we practice at the small scale sets the patterns for the whole system."[47] There was a fractal in the free maroons of Saint-Domingue who stormed down from their mountains into combat with French troops so that the whole island could be free. There was a fractal in the spinning wheel on the Indian National Congress flag, extending from a traditional practice to an eventual industrial policy. Resistance can spread up and down the stack. Carefully chosen practices sever habits of dependency on the systems that otherwise seem inevitable. Echoing the Cold War–era Non-Aligned Movement among countries caught between the United States and the Soviet Union, governable stacks could be the basis of a new movement of digital non-alignment, asserting many diverse sovereignties against the dueling forces of Silicon Valley and Shenzhen.[48]

Stack Design and Pedagogy

Not all intentional stacks are governable. Groups dedicated to racism and authoritarianism have become particularly intentional about their network stacks, migrating to dedicated platforms such as Stormfront, Parler, and Gab as more mainstream networks remove them. These have tended to build their communities more around the appeals of persecution and provocation than promises of self-governance—although Parler, for instance, pioneered user juries for enforcing its sparse content-moderation policies.[49] Stacks are contestable spaces, and some self-governing is no guarantee that anything good will come of it. The particulars of design matter immensely, as do the kinds of political skills that communities teach each other.

For any layer or component of a stack, we might ask a common set of questions, along three vectors:[50]

- *Sovereignty*: Who is ultimately in control, and how? Is there too much reliance on external resources? What happens to the value that derives from labor and culture? How easy is it for individuals and communities to exit if they so choose?

- *Democracy*: How can participants be part of the flows of power? Are those flows explicitly stated and widely understood? Are interfaces accessible and culturally appropriate?
- *Liberation*: Does the stack resist systems of exploitation? Is it centering people and experiences that other stacks marginalize? Does it reduce unwanted dependencies? How could it spread to other communities and make self-governing easier?

The point of these questions is not a litmus test for knowing what is or isn't a governable stack. The point isn't to achieve governability and be done, but to continually seek more of it across more layers and vectors. The stack is never complete, any more than a community can be. Sometimes governability is possible through reconfiguring tools already available, or perhaps it is necessary to make new ones. Tiziana Terranova, who has proposed the idea of a "red stack," writes that insurgents can build "new platforms through a crafty bricolage of existing technologies, the enactment of new subjectivities through a *détournement* of widespread social media literacy."[51] One way or another, the point is to organize technologies that can bend with the ungovernable contortions of self-governing— technology for communities that can be, as Rosa Luxemburg hoped for, "supple as well as firm."

In the sense of Grace Lee Boggs's dialectical humanism, governable stacks invite the people who use them to change their relationship with technologies, to imagine different sorts of technologies, and to be changed themselves. We learn with each other, and we learn with the machines, which take on life of their own. Governable stacks enable what Christopher Kelty calls "recursive publics"— communities whose work is, at least in part, the making of what makes their community possible.[52] The stack is a cyborg cycle, and it is pedagogy. Crafting it, across its layers and vectors, means learning with it.

The Detroit Community Technology Project, developed under the tutelage of Grace Lee Boggs, uses education through stacks as a strategy for self-governance. The organization trains people to deploy locally managed internet infrastructures, particularly in majority Black neighborhoods that have been systemically underserved. In this work, organizers refer to Boggs's maxim of stressing "critical connections" over "critical mass."[53] This is because setting up a local WiFi node on an apartment building may seem small compared to the scale of a regional telecom monopoly. But in the shaping of imagination for people involved, small interventions like this can do far more than the scalability of the telecom ever could. To shift the stack and to learn with it is to make a rupture. While a stack run from above provides mere service, a governable stack can introduce experiences of shared power. Those experiences can shatter the telecom's claim that its dominance is inevitable. Whoever touches the governable stack risks recognizing that another kind of relationship with technology is possible.

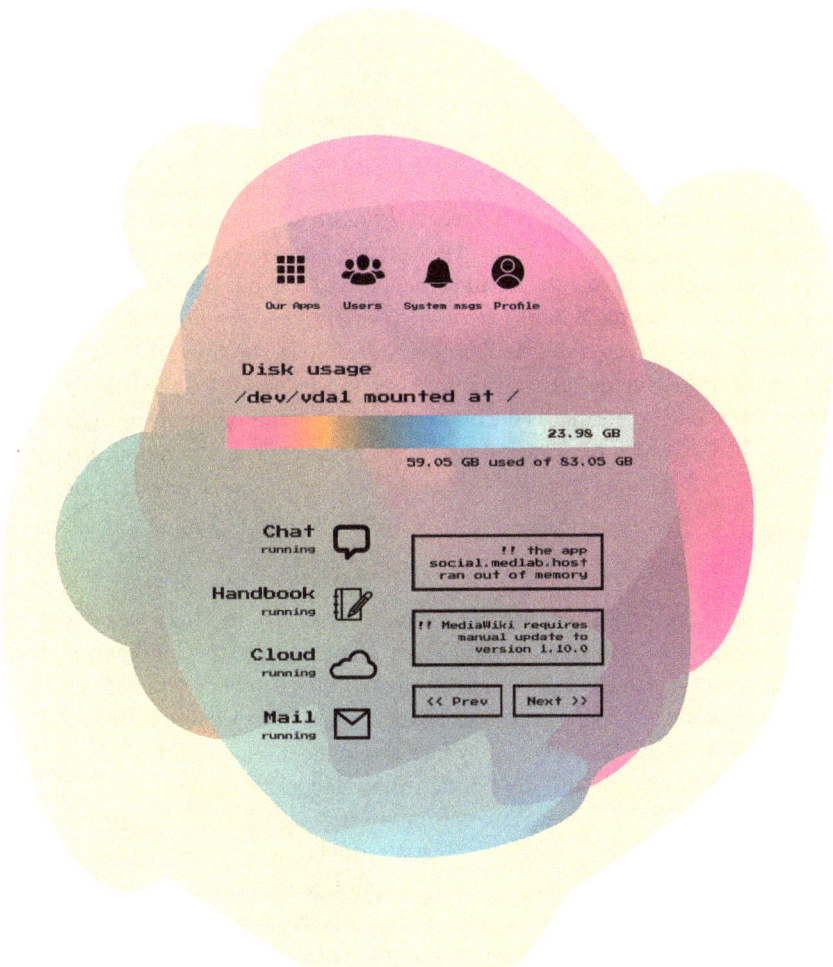

FIGURE 12.

The university lab I direct has also sought to manage a stack as an exercise in pedagogy and space-making. We operate our own suite of software for chat, file sharing, polls, websites, and multiplayer games. Students use these to collaborate, and those who are interested can learn to be co-administrators.

The lab's "cloud" is an ever-evolving experiment, still short of what I would hope for from a governable stack. Most students have yet to play much of a role in decision-making or design. The stack also resides on the servers of a faraway hosting company; I hope someday that students can hear the hum of the machines running their tools. And I question whether our stack is challenging any colonizers. Its hiccups often seem to remind students why they prefer systems that powerful companies manage for the price of their data. At least so far, I fear that our steps toward governability might have taken us at least as many steps back. We feel alone in what we are doing, and that makes the frustrations all the more demoralizing. It becomes easier for any one group to make its stack governable when others are doing the same.

What user-experience designers call "friction"—when a technology requires extra work from users—is instructive. Friction reveals what is not being designed for and what runs against the grain of dominant systems. Friction happens a lot for those building governable stacks, and it happens a lot in our lab. But then there are also times when the stack simply works, to the point that we stop noticing all of the tinkering and learning that it took to get there. Governability feels available and obvious. These moments are worth observing, too, because they show that governable stacks could be normal as the organizing logic of our online lives, so expected and obvious that we have to stop ourselves to notice. When we do notice, we start to see how rudimentary governability could become the basis for even more.

MODULAR POLITICS

In my town there is a manufactured-home park that has been searching for the right technology to support its self-governance.[54] It recently became a cooperative when the residents organized to buy out their landlord. They are sensitive to the prospect of ending up in an exploitative relationship again. As they explore what their stack might involve, they face a minefield. The local telecoms have a history of poor service and high costs in low-income communities. Corporate cloud services for file-sharing and communication aren't well equipped to serve residents who, in many cases, lack access to the latest machines and apps. Popular collaboration software does not have features meant for cooperative decision-making. Implicit feudalism reigns. Every layer of the stack grinds against their self-governance—a burden that the residents don't have time or money to deal with.

Imagine, then, a different set of options instead. Internet service comes from a local cooperative, deploying high-speed fiber connectivity at cost; one of the residents is on the company's board. Along with similar communities elsewhere, the residents are part of a software cooperative that provides communication tools focused on self-governance among people with varying access to devices. The major processes outlined in their bylaws occur on the platform. After a few years, the residents decide to shift from having a single board to organizing through working groups, each focused on particular aspects of running the neighborhood. On their platform, they simply replace the Board plugin with one for interconnected Circles. When some members fear the platform is collecting too much personal information about them, they are able to satisfy the concern with a discussion at the platform's next annual meeting, where they pass a resolution that changes its data retention policy.

Elinor Ostrom conceived of the experience of self-governance as an "action situation." Faced with a decision, what choices does a person or group have at hand? An action situation occurs within an "action arena," the context that situates the available options. What changed for the mobile home park between the reality and the speculation was how a different stack makes for a different action arena.

Modular politics is a model for the design of action arenas in online spaces that I developed with my collaborators in the Metagovernance Project, an online network of researchers and builders.[55] We imagine this model as a foil to implicit feudalism, the basis of an emerging "governance layer for the internet." It is a workshop for artisans of self-governance. To that end, we outlined four design goals:

Modularity	Platform operators and community members should have the ability to construct systems by creating, importing, and arranging composable parts together as a coherent whole.
Expressiveness	The governance layer should be able to implement as wide a range of processes as possible.
Portability	Governance tools developed for one platform should be portable to another platform for reuse and adaptation.
Interoperability	Governance systems operating on different platforms and protocols should have the ability to interact with each other, sharing data and influencing each other's processes.[56]

Together, these goals provide the foundation for experimentation with and the circulation of governance designs—exactly what implicit feudalism inhibits. Tools that implement modular politics could be embedded in many kinds of online spaces, from social media and productivity tools to labor markets and virtual classrooms. Modularity means that insights from one kind of community can be combined with those from another. Portability means that a third community can adopt them both, even in a different kind of technical and social context. A group of environmental activists, for instance, could adopt a voting module designed for an online game and connect it with their own code of conduct. Interoperability means that the group's decisions could spread to other similar groups around the world; when a critical mass of them agree about something, it could trigger a global mobilization. Expressiveness means that modules can be designed to enact many kinds of processes, reflecting diverse cultural traditions and regional norms. Evolution thrives on diversity. No system will be neutral, but designers can set out to make it as pliable as possible, avoiding the temptation to simply replicate the architecture of the computer or the culture of its builders in the design of social spaces.

In 2017 I was part of a small group that founded Social.coop, a self-governing social network. Our primary service for members is to maintain a server running

Mastodon, the open-source social media platform.[57] To set up the system in a way that resembled even the most basic kind of cooperative, however, we needed a lot more than Mastodon, whose design cleaves to implicit feudalism. For deliberation and decision-making, we turned to Loomio, the platform developed on the model of consensus process in the 2011 Occupy encampments. To clarify what counts as a decision and how to hold one, we set up a wiki to manage bylaws and other documents. The payments platform Open Collective, designed to support open-source projects, enabled us to collect dues, pay expenses, and manage our membership. Working groups turned to Matrix chat rooms for day-to-day operations. In order to assemble a governable stack, we had to make the internet bend over backward and require our members to create way too many accounts. Even then, our self-governance has continued to feel like a necessary hack, like we are always paddling upstream rather than following a natural flow.

If platforms like Mastodon were to support a modular framework for governance design, stacks could evolve more in step with the communities that use them. While some layers of a stack should serve as a stable foundation, others might need more rapid experimentation—just as national constitutions are harder to change than local laws.[58] The increasingly divergent rulesets in different language editions of Wikipedia, for instance, suggest that online communities can benefit from adapting their governance to particular contexts.[59] As Elinor Ostrom put it, evolution across multiple communities helps produce institutional diversity: the mix of overlapping, interacting structures that reflect the complex realities and needs of human societies.[60] Modular designs can enable stacks to better reflect the multiplicity of their communities.

Since developing the modular politics framework, my collaborators and I have begun to see it coming to life. One of us, Amy X. Zhang, has developed a prototype governance platform called PolicyKit, which adds governance functionality to popular social platforms; the Metagovernance Project experimented with making it more modular and expressive through a further prototype called Gateway.[61] At Social.coop, we have used Gateway to integrate our cash flow on Open Collective with our decision-making on Loomio; once a decision reaches a certain threshold of approvals, the payment attached to it proceeds automatically. In small ways like this, we have begun to experience inklings of a governable stack.

Modular thinking has been spreading far more widely than our experiments. The civic participation platform Decidim, used largely by city governments for citizen feedback, has a modular structure. Its growing library of modules ranges from specific decision-making mechanisms to integrations with other platforms.[62] The platform continues to evolve through a governance process that runs on the platform itself. But most explorations of the modular approach have been in the context of blockchains—the kinds of online spaces where shared ownership is the default, where co-governance of some kind is necessary for anything else to work.

The stacks that support DAOs and other crypto communities need to include at least some basic technologies for participant governance. Safe, the most popular "multisig" wallet that DAOs use to manage their digital assets, expects communities to set a certain threshold among their members to approve a transaction. But Safe also supports a project called Zodiac, "an expansion pack for DAOs" that enables communities to create and adopt diverse governance modules. Another widely used tool for building DAOs, Aragon, has been entirely rebuilt with a modular design, supporting governance applications that run on a core "kernel." OpenZeppelin, a software library for building crypto applications enables users to design and assemble governance processes with modular bits of code.[63] My collaborators at the Metagovernance Project have convened these organizations and more like them in DAOstar, an effort to develop shared standards for DAOs, enabling greater portability and interoperability among them.

Systems that implement modular politics offer a wider canvas for governable stacks. The canvas raises a new set of questions: What palette will people use to paint it? What habits and biases and histories will inform the images we create?

GOVERNANCE ARCHAEOLOGY

Cowrie shells may be the most widespread and persistent kind of money in human history. The former homes of small mollusks, the shells are usually smooth, even shiny, except for the toothed edges that run along a lengthwise slit. For millennia they have been used to store and exchange value from Africa, China, and India to inland parts of pre-Columbian North America. Europeans harvested them in bulk from the Indian Ocean in order to buy enslaved West Africans. But they were not just cash. Cowries have also served as jewelry, aids in divination rituals, gambling chips, and ballast for ships. On the wampum belts of Indigenous North Americans, they served to establish contracts, treaties, and histories.[64]

From the financial to the mystical to the artistic, the cowrie's array of uses is not unlike what people hope to enable with blockchains. This new kind of programmable ledger may not be as wholly new as some claim. Among the creative and horrific annals of cowrie use, surely there are lessons for making governable stacks today.

How we imagine governance histories will orient our responses to governance crises and governance opportunities. "When people decide important matters, they turn to the past," writes political theorist Anne Norton. "They look to history and custom, they consult the advice, the wisdom and the dreams of the past. They are not bound by the past, but they bear it in mind. The past does not rule them, but they go forward mindful of those who came before."[65]

Thomas Jefferson's library, now reconstructed at the Library of Congress in Washington, DC, reflects a culture concerned with mimicking Greek and Roman antiquity. The founding governance documents he co-authored root their authority in that particular history. The Indigenous societies of his immediate surroundings also influenced Jefferson and his ilk, but his colonial ambitions depended on regarding them as "savages," not as sources of inspiration.[66] Organizing a new institutional order is in part a matter of organizing a set of relationships with one's predecessors.

Governance archaeology is a practice of intentionally crafting relationships between new governance designs and preexisting legacies. Conventional archaeology deals with the remnants of the past that are still here in the present, unearthing them for study and displaying them for the purposes of the living; governance designers do something similar, whether consciously or not. They draw on their muscle memory and their ancestors when deciding what seems right and what might work.

Political scientist Federica Carugati and I began devising the concept of governance archaeology as we assembled a database of collective-governance institutions across time and space.[67] Our hope for the database was to find ideas applicable to present challenges in the online economy, but its applications extend beyond just online contexts. If political institutions are ripe for reinvention all around us, what kind of library will inform their replacements?

Recent popular works of "big history" attempt to render the long sweep of the human past useful for innovators. Yuval Noah Harari's *Sapiens*, widely read in Silicon Valley and its allied subcultures, regards technology as an especially motive force, constraining and unlocking the spiritual-social options of any given epoch. In response, David Graeber and David Wengrow's *The Dawn of Everything* retraces the archaeological record as a story of staggering diversity in governance forms, an invitation to devise similarly diverse arrangements in the present.[68] Both works have captured public attention and appear on the bookshelves of today's elites. Governance archaeology is an attempt to make the relationships between legacies of the past and designers of the present more explicit, more rigorous, and more self-aware. The goal is not simply to amass a larger quantity of reference points but to refer to them more responsibly.

The case of Jefferson is a reminder that colonial relations distort historical knowledge—from his nostalgic perception of southern Europe to his erasure of the Indigenous federations and the African diaspora around him. Governance archaeology must see such power relations and interrogate them. A decolonial posture might begin with two steps: expanding the canon of democratic legacies while repairing relationships with legacies that have suffered violence, ignorance, and subjugation. On repeat, these open us to what decolonial theorist Catherine Walsh describes as "a past capable of renovating the future."[69]

To expand the canon is to attempt something like the "ecology of knowledges" that Boaventura de Sousa Santos proposes.[70] In such an ecology, cowrie shells and blockchains can inhabit a common universe, together with the coins of medieval Italian city-states and the concurrent *hawala* money-transfer system across Islamic trading networks. Among these, de Sousa Santos challenges us to practice "radical copresence": a juxtaposition across lines of culture and power that refutes the centrality of the dominant narratives. For instance, Athens was but one example of democratic governance in the ancient world. Republics could be found among cities in what is now India, including cases of choosing leaders by random lot. Hereditary chiefs around the world have had to respect long-evolved collective decision-making processes in their communities.[71] Each social artifact we collect in our database is distinct, and each bears lessons that could inform the design of governable stacks. The Western canon of political history becomes only one legacy among many.

The second step of governance archaeology, the repair, means cultivating relationality. It aspires toward ancestry—learning to regard those we learn from as political ancestors, while we work to become good descendants. "The role of the ancestors," explains Ronaldo Vázquez in an essay on decolonial listening, "is not a passive or a conservative one, but rather an active source of meaning."[72] Descendants should want to be worthy of what they learn. They must also be willing to question their ancestors' convictions and add their own experience to what they inherit.

Transformative justice activists, for instance, frequently acknowledge that practices such as accountability circles draw on living-yet-suppressed Indigenous legacies. Through the adoption of those practices, alliances form. They use the term BIPOC—Black, Indigenous, and people of color—to stress solidarity between the two most violently oppressed groups in US history. They recognize efforts to address assault in Black communities alongside struggles seeking justice for missing and murdered Indigenous people.[73] Common practices breed common causes.

Ancestry is not a one-way relationship. It is not automatic. It asks more of designers than to take and apply; it expects reciprocity, and reciprocity comes with opportunities of its own. Perhaps, before including a historical voting mechanism on a governable stack, designers should speak with the direct descendants of the people who developed that process and ask how they see it today. Asking permission may be appropriate if there is an authority in a position to grant or refuse it. When a stack produces value from a community's insights, royalties or reparations might go back to that community. There is no formula for reciprocity. Yet if the current moment is to be a formative one, akin to that of Jefferson, the new governable stacks should relate to their precursors more honorably than he did.

Stacks are assemblages of living beings, institutions, and technologies, assembled so the components can be more powerful together. It is for power, also, that militaries and corporations assemble stacks under their own control. Colonial stacks are ubiquitous in online life for many of us. They impose surveillance, economic exploitation, and social control within and across borders. Long before digital colonialism, the anticolonial tradition has shaken off empires through techniques and technologies of self-governance. The act of making useful, governable stacks will refute colonial claims that democracy has no place on networks or that it is too difficult. Governable stacks are confrontations. They wear down the reigning assumptions. They show how so much more of our online world could become governable space.

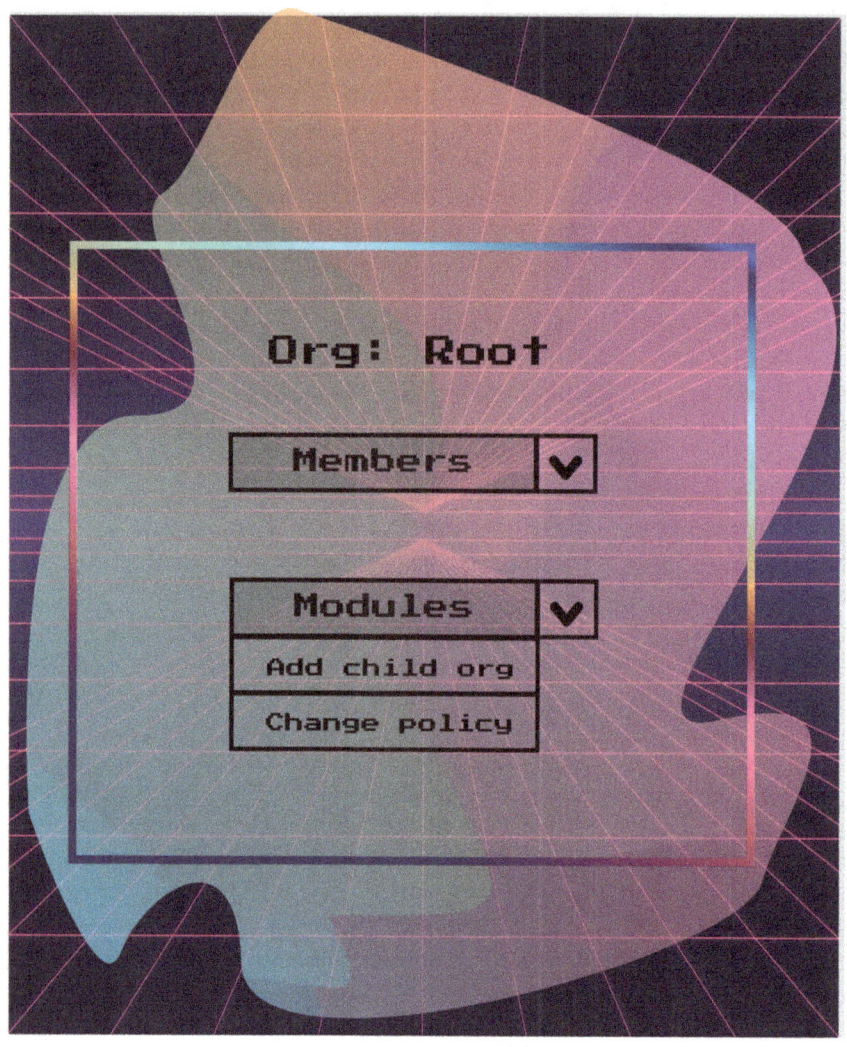

FIGURE 13.

Modpol

modpol.net

Modpol is a self-governance toolkit for communities in online worlds. My collaborators and I created the first implementation in a multiplayer game called Minetest, an open-source, noncommercial game developed by its players. Minetest resembles the more popular Microsoft-owned game Minecraft. Our goal was to translate the modular politics framework described in chapter 4 into code. Doing so has forced us to clarify the framework in greater detail than outlining it in words and to contend more directly with our underlying assumptions and biases.

With Modpol, Minetest players can form groups, called "orgs," and choose the set of governance modules available in the orgs they form. They can also create their own modules in Lua, a programming language often used for modifying games. Modules can activate other modules; a module to admit a new org member might call a module that needs everyone's consent, or it could call a coin-flipping module, or it could defer the question to another org. While figuring out how to make this work, there were a few design decisions we made that helped Modpol depart from the pattern of implicit feudalism:

- *Groups over roles.* Instead of assigning powers to particular users, Modpol assigns powers to orgs. Ultimately, it is on the level of org membership not individual permissions that things happen. Orgs can make decisions using whatever modules they choose. They can use the consent module we created to approve decisions with a certain threshold of votes; they can also defer an approval to a one-member org if they want a role-like structure. But sovereignty stems first from the collective, not an individual.

- *Freedom over authority.* The default setting for org decisions is trust—any user can take any available action within an org. The system does not assume that one admin holds all the power. Org members can change that and create an admin structure (or anything else), but they have to do so intentionally. Autocracy is just one option among many for how to run a group, rather than being the presumption at the outset.

- *Inheritance over blank slates.* Usually, new spaces for online groups on a platform start out the same. Real social life, however, is infused with habit, tradition, and muscle memory. Modpol reflects that. When new orgs form within existing orgs, they inherit the rules of their parents. Those rules can be changed. But the rules begin with whatever users were already doing.

Minetest is a game for building worlds. Players explore landscapes, gather resources, and use them to create the kinds of spaces they want to inhabit and show off. Modpol is also meant for building. Players can create worlds of interlocking orgs, each with their own rules and processes. Modpol could be used to organize teams for Capture the Flag or to govern an anarchist castle. It is an engine for organizing self-governance.

5

Governable Spaces

Democracy as a Policy Strategy

"Monsters Come Howling in Their Season" is a story set a few decades in the future on the Caribbean island of St. Thomas, where the author, Cadwell Turnbull, grew up.[1] The monsters are hurricanes. One of the characters speaks the words of the title to register his acceptance of the ever more frequent and ferocious storms due to climate change in the mid-twenty-first century. They come, but they no longer devastate. Carbon emissions have become negligible. This is due to the story's central figure: a computer system called Common, whose access to the intimate details of residents' lives enables it to coordinate their actions to protect themselves and their neighbors from the storms. People share their lives with Common, we learn, because they co-own it. Common is governable. When people get anxious about Common's presence, they can decide together how to program it differently. Turnbull explains: "Common is governed by a federation of collective institutions from all over the world that are devoted to the mission of AI as a public resource. Anyone can add knowledge to Common, and there is a democratic process to building the hardware necessary to carry the AI. Common is decentralized and spread across all of the devices that run its software. Tech cooperatives create vessels to hold the AI—from literal black boxes to giant robots—but most people use practical vessels like smartphones and watches."

The result of these overlapping structures is trust. At the end of the story, Common asks the narrator, "Can I remember this conversation?" The narrator recalls, "I consider the question for a long time. Then I shrug. 'Okay.'"

Adrienne maree brown has often turned to science fiction as a practice of social change. "We hold so many worlds inside us," she writes in *Octavia's Brood*, a collection of stories by activists, an homage to the fiction of Octavia Butler.[2]

"It is our radical responsibility to share these worlds, to plant them in the soil of our society as seeds for the type of justice we want and need." She encourages the practice of "science fictional behavior." Elsewhere she adds, "I believe that all organizing is science fiction."[3]

Brown says this in the context of a genre that has often been White- and male-dominated, reflecting a hierarchy of whom society has invited to imagine and create its future. In Turnbull's story a character asks, "Did you know we were one of the first to use Common for hurricanes?"—not Silicon Valley or MIT, that is, but the residents of a Black-majority island on the front lines of climate change. It is audacious to imagine an artificially intelligent system accountable to the people of St. Thomas, considering how such systems today are most often used to manage people on behalf of corporations and governments.[4] It takes science fiction to begin imagining a different economy of innovation, a different distribution of opportunities and rewards for instantiating the new. The same might be said for imagining a different way of making the policies that structure online life.

This is a chapter about policy. In what follows I formulate a strategy for policy design based on cultivating *governable spaces*. Governable spaces arise when social and technical infrastructures enable participants to deliberate, make decisions, and enact those decisions through accessible, transparent, and just processes. To the extent that systems of rules organize our societies, governable spaces are difficult to achieve without policies that are well suited for supporting them. This chapter will explore how to develop policy that supports online self-governance. As with governable stacks, however, I will not provide a list of minimum conditions for what is or isn't a governable space. Governable spaces are a vector, a direction of motion, not a standard or condition that can be named without knowing its context.

First, I will show how governable spaces can be sites of problem-solving for vexing challenges in three domains of the online economy: social-media communities, platform-mediated work, and network infrastructure. Then I identify arenas of policy that could help enable governable spaces to take hold more widely: governments, organizations, and technologies.

Toward those ends, I find that feminist tradition is especially instructive in its insistence on holding space and time for self-governance and in making room for people to bring their whole selves into it. I also draw on my years of studying and supporting cooperative startups in the online economy. For generations, cooperativists have demanded what has come to be called a "partner state"—public policy that encourages democratic associations across society without attempting to control them.[5]

I veer the end of this book into policy not because it is the destination toward which all else leads, the realm of ultimate importance. The future of democracy does not lie simply with what laws do or do not pass in governments. Rather, I mean to show how policy can enlarge the reach of foregoing concepts like political skills, democratic play, and governable stacks. The policies I am interested in, also, do not come solely from legislators and regulators. Already in the online economy, territorial governments are not the only governors; policies of comparable effect also appear in the code of platforms and protocols and in the text of corporate structures and employment contracts.[6] Changing how we design policy can also mean changing where policy appears.

Certain assumptions about policymaking tend to prevail around the reigning online spaces. When problems of human behavior arise, users and governments alike expect the companies that run the platforms to take charge and enact solutions. The solutions need not be transparent or accountable as long as they occur. This expectation stems from a further assumption that when complex technology is involved, social problems are best understood as engineering problems. Because the platform companies have plentiful engineers, they are evidently best suited to solving the social problems that arise on their technologies. When the companies cannot engineer a social problem away, the thinking goes, there is need for a higher authority to take charge, such as the fiat of a government regulator. Each of these assumptions makes a certain kind of sense, but here I point to an option they ignore: problem-solving through self-governance.

The heart of my argument is a call for shifting the orientation of policymaking from top-down regulation, which reinforces existing sites of power, toward enabling new sites of power among user communities. This kind of policy seeks to ensure that people have the power to solve problems on their own terms. In the process, policy can secure a future for democracy by letting it evolve—under diverse conditions, confronting urgent needs.

Seeding governable spaces through policy involves work akin to what brown undertook to assemble *Octavia's Brood*: inviting activists from marginalized communities to write science fiction, a genre that has so often left them out of its futures. How could co-governance come within reach of everyone? If we can answer that question, then societies like Turnbull's St. Thomas—leading the world from the margins—might become thinkable and achievable.

PROBLEM-SOLVING WITH SELF-GOVERNANCE

Much of the idealism around internet regulation has aspired to produce a network that is open, neutral, and universal. Metaphors like "net neutrality," "global village," and "platform" itself all reflect that aspiration.[7] But a recurrent theme in feminist critiques of internet culture, as well as in feminist thought more generally, is suspicion toward allegedly neutral forms of organization.[8] This suspicion comes from experience. Female-presenting social-media users, especially those with intersecting marginalized identities, face disproportionate hostility and exploitation due to policies that claim to support free speech. The algorithmic labor management of gig platforms has reinforced segregation and subjugation in labor markets.[9]

Critiques of facial recognition, surveillance, and online search demonstrate that services designed for "anyone" may in fact do harm to people whose bodies and experiences are not those of the designers and investors.[10] Intersectional feminism has seen with particular acuity how the online economy has contrived to be both apparently neutral and persistently unaccountable. Government regulators have meanwhile embraced the platform companies' claims of being neutral infrastructures, while conferring on them both the power and responsibility to govern from the top down.[11]

Women and feminist perspectives played a significant, overlooked role in the early development of computing cultures. Feminists have since extended earlier analyses of undervalued labor such as housework to digital worksites, from system engineering to the emotional labor of community management.[12] This legacy brings us back to what Jo Freeman observed among early-1970s feminist "rap groups" in her famous essay against "The Tyranny of Structurelessness": that groups lacking clear processes or pathways for participation often fall into rigid hierarchies and stewing resentment.[13] "Those who do not know the rules and are not chosen for initiation must remain in confusion," she wrote, "or suffer from paranoid delusions that something is happening of which they are not quite aware." In response, Freeman offered proposals for "democratic structuring"—practices such rotating roles of authority, ensuring that power-holders are responsible to the entire group, and diffusing information widely.

More recent feminists have similarly seen fit to establish intentionally bounded gathering spaces, along lines of gender identity, racial politics, and affinity; within these spaces, participants often develop clear codes of conduct.[14] Feminist media scholars have further called for participation and community control as a means of transforming media environments that have historically marginalized them.[15] A Malaysia-based international process to produce feminist principles of the internet included in its final document a call to "democratise policy making affecting the internet as well as diffuse ownership of and power in global and local networks."[16] The pursuit of governable spaces is a strategy for policy that takes the need for democratic structuring seriously.

The feminism considered here includes a range of sources, not necessarily ones belonging to a single wave, strand, or lineage. Surely there are feminists who would challenge the tendencies I highlight. I am not seeking to alter or summarize feminist tradition but to identify patterns that it has seen especially clearly. Freeman's essay serves as a gravity well that attracts shared concerns among diverse feminist perspectives. Together, these perspectives reveal ways in which male-dominated technology companies have leveraged patriarchal relations into structures of top-down control. Patriarchy thus reconfigures itself as paternalism through allegedly beneficent entrepreneurship—the "exertion of positive rather than coercive power," as Liena Gurevich describes the paternalist impulse.[17] The prevailing discourse among online platforms tends to take paternalist rule for granted as necessary and legitimate; feminist tradition has far less patience for doing so.

This section extrapolates from feminist scholars' attention to self-governance by outlining a strategy for governable spaces across various domains of controversy and policymaking in the online economy. The upshot of each exercise is to reconfigure supposedly neutral regimes, frequently managed through monopoly power, with self-governance and accountability. Doing so involves new forms of boundary-making and rule-setting against the ambitions of corporate and regulatory monocultures. The strategy I propose seeks not rigid central planning but lived environments crafted collectively over time.

Governable Communities

During the waning days of 2020, US president Donald Trump threatened to veto the annual National Defense Authorization Act if it did not include a provision unrelated to national defense: a reversal of Section 230 of the 1996 Communications Decency Act. This was one more instance of melodrama in the career of a snippet of law that has become known as "the twenty-six words that created the internet."[18] For speech that occurs on an "interactive computer service," it pins liability not on the service but on the user. Doing so frees online platforms from most responsibility for what users publish on them, making possible an industry based on user-generated content. The discontent comes from two opposing directions. Some critics denounce Section 230 for enabling social media to become a cesspool of hate speech and disinformation, while others—Trump, for instance—accuse the law of freeing platform companies to partake in arbitrary or partisan censorship.

Despite its reputation as a permission slip for online disorder, Section 230 cleared the way a new order of governance. This regime has spread far beyond the United States with the global influence of Silicon Valley platforms. The legislators who wrote the provision hoped their safe harbor would encourage services on the fledgling internet to self-regulate without fear of political meddling.[19] Platform companies thereby became what Kate Klonick has called "the new governors"—not merely moderators or enforcers but architects of meaningfully autonomous polities wielding power over users and the public sphere. Internationally, the assortment of governors is becoming ever more complex, straying far from the dream of a global village. The internet appears to be drifting toward a "splinternet" of conflicting regimes, requiring global platforms to behave differently among jurisdictions, if they are allowed to operate at all.[20]

From the perspective of most social-media users, content moderation is a matter of imposition, whether by remote company owners or by the more proximate volunteer administrators. The design pattern of implicit feudalism relies on power-holders who are not chosen or removable by those they govern. Rule enforcement occurs through censorship of user content or the removal of users altogether, but rules do not necessarily apply to the administrators themselves. Users can speak out or leave online spaces, but they lack the direct levers of effective voice. This contributes to the "techlash" against platform companies that spreads with every scandal of content moderation and abuse; by hoarding power, the companies have hoarded the blame.

Given the centrality of Section 230 to the online economy, politicians' calls to eliminate it may be mere posturing. One of the more constructive proposals for reforming Section 230 would remove protection from platforms that act as "bad Samaritans" by actively encouraging toxic or criminal content.[21] But this proposal and others like it still presume a platform-centric approach to content policymaking, rather than one centered on the governance that user communities might conduct for themselves. The company-customer relationship so central to industrial markets remains the preferred logic of regulation, not the peer-to-peer relations that prevail in the lived experience of online life. For instance, the European Union's General Data Protection Regulation expects privacy rights to accrue from the actions of company bureaucracies and unusually zealous users; the potential for collective action is only beginning to be explored.[22] But what if network polices better reflected the experience of networked relationships?

Feminist political theorists have retrieved and radicalized the ancient recognition, articulated by the likes of Confucius and Aristotle, that healthy governance grows from the ground of friendship among citizens. Aristotle observed friendship as having the power "to hold states together." Although no great admirer of democracy, he found that friendships "exist more fully" in democracies than in other systems. For Confucius, friendship was the only one of the five relationships basic to a healthy society that does not depend on hierarchy.[23] To practice equality in everyday life is practice for governing; modern feminists have gone further to see friendship as a basis for evolving the social order through self-governance from below.

For example, Marilyn Friedman has argued for centering peer relationships, rather than the patriarchy-inflected family or territorial community, as the starting point for liberatory politics. In friendship she sees the basis of communities able to support an embrace the "idiosyncratic" and "unconventional." "Friendship," Friedman writes, "has socially disruptive possibilities." It can be the basis of a feminist communitarianism—community by mutual volition rather than by accidents of birth.[24] But social media platforms constrain what friendship can do as a basis of social and political power. Even while platforms have opened new opportunities for friendships among peers, instrumental power flows from company CEOs down to the feudal admins and mods, according to terms that government regulators set or fail to set.

One step toward making community spaces more governable is for users to establish clearer boundaries and purposes where they gather—echoing Virginia Woolf, spaces of our own.[25] Legal regimes might expect subsidiarity, as discussed in chapter 3, as a prerequisite for protection from liability. Rather than encouraging scalable governance by platform companies, the law could expect user self-governance at the scale of communities. Platforms would gain immunity only by sharing power.

Under policy that expects governable spaces, social networks would have incentives to design for healthy self-governance. They would have to provide for users something on the order of modular politics—tools that support a variety of participatory mechanisms for rule-making and administration, such as elections, petitions, boards, and juries. Rather than relying on assignments of permissions to individual users, default settings might assume decision-making as a collective affair. For instance, the European food-sharing platform Karrot allows a local community to remove a member only through a group process, rather than by the fiat of a single administrator. While such an arrangement lies well outside the norms of social-media systems, it is at home in feminist conceptions of the relational self, the person as a nexus of relationships.[26]

Online life has already taught us that satisfying everyone with universal rules from above is doomed to fail. If social platforms became regulated on the premise of self-governance, the responsibility for what happens on them could be more sensibly shared.

Governable Work

Before she was a legal scholar, Sanjukta Paul had a job at a labor union. She saw how US antitrust laws—ostensibly intended to constrain corporate power—actually narrowed the options workers had for joining together and organizing. Policy, she came to realize, acts as an "allocator of coordination rights": an arbiter of who is allowed to team up and how.[27] While the constraints of US antitrust doctrine on labor organizing are specific to the country and context, law everywhere plays this role of allocation.

Restrictions on coordination can be difficult to notice, lurking in the shadows of what the law prevents, even without actively prohibiting it. Feminist scholars have chronicled how the policing of coordination has been a means of undermining women's collective power, from the persecution of witchcraft to the isolation of suburban homemaking.[28] Just as witch hunts sought to keep women's economic lives dependent on patriarchal relationships, laws today help preserve a fragmented, atomized workforce available for gig platforms and other precarious jobs. While antitrust law is only rarely wielded against large platform companies, in many countries it imposes legal barriers that have prevented platform workers from forming unions or cooperatives.[29] Paul invites us to ask who is and isn't allowed to find common cause.

Economist Juliet Schor's *After the Gig* presents the story of the platform-mediated gig economy as actually many stories at once.[30] Schor draws on close-up studies of platform workers—the drivers, the deliverers, the hosts, the doers of various tasks—and reveals their cleavages. Some workers find a kind of liberation, while others fall into a trap.

Schor and her research team constructed a kind of missing conversation through their interviews across a field of dispersed experience. Unlike social-media platforms, gig apps discourage persistent relationships among users, whether they hold worker or customer roles. The platform claims to supplant the need for relationships. Such user-experience designs, like early-twentieth-century US antitrust laws, establish policies for coordination rights. The platforms organize those rights on behalf of managerial control. Users get apparently open and frictionless transactions but no durable means of seeing each other, of comparing experiences, of finding the wherewithal to co-govern.

Feminists have long sought to reveal and recognize the significance of work that patriarchy would prefer to keep invisible and underpaid.[31] Before online gig platforms, women performed piecework for the textile industry under similarly precarious regimes; the precarity continued when women seeded the computer industry by doing rote computation and early programming—only to be discarded when they had sufficiently trained machines to take up their work.[32] Because feminist tradition has been attentive to these otherwise neglected histories, it bears conceptual tools well-equipped for the present abuses often euphemized as "the future of work."

Examples are widespread. Emotional labor and reproductive labor enable the digital economy to function, while the credit for production typically goes toward technical systems and male entrepreneurs.[33] Tech companies increasingly depend on little-seen and poorly rewarded "ghost work" that occurs in homes or offices far from the tech hubs.[34] If a social-media company succeeds in removing violent imagery from its platform, is that because of the executives' policies and the engineers' algorithms or the offshore workers who have to look at things all day they will never be able to unsee? Workers' unpaid family members organize meals and schedules that make the paid work possible. Feats of governance happen not just in executive boardrooms or shareholder votes, not just in algorithms and user-experience, but in the daily negotiations that companies intentionally hide from view.

Schor holds out hope for the possibility of freer, less wasteful, platform-mediated future of work. But "achieving the potential of platforms requires specific conditions," she writes.[35] In particular, she highlights efforts to develop cooperative platforms, owned and governed by their workers.

Ra Criscitiello, the deputy director of research for Service Employees International Union–United Healthcare Workers West (SEIU-UHW) in California, attempted to create a gig platform that her union's members would co-own. NursesCan, as they planned to call it, would connect patients and workers for at-home, on-demand healthcare services. But building a viable cooperative in a tech economy made for investor ownership and venture capital did not prove feasible; investor-backed competitors had access to far greater resources. Criscitiello responded by becoming more ambitious.

In the storm of California's struggles to define the policy environment for gig work, she initiated a state-level proposal called the Cooperative Economy Act, a version of which was introduced to the state legislature in 2021.[36] The bill proposes a federation of tax-advantaged, employee-owned cooperatives that could contract with online labor platforms. Workers could thereby collectively bargain over the terms of their work for platforms, without the platforms needing to employ the workers directly. Workers would elect their co-ops' leaders. Although California's 2020 law known as Proposition 22 exempted platform drivers and delivery workers from the rights associated with employment, other platform workers—such as SEIU-UHW's nurses—stand to benefit from organizing their gig work as employees. The proposal shares features with the Hollywood system, devised long before the internet, which enables the film production workforce to move from job to job while retaining union representation and sharing in the studios' profits.[37]

Even in the absence of legal cooperatives or unions, workers are finding ways to gain fuller control over their livelihoods. Platform-based drivers in Colombian cities, for instance, use group chats and other technologies to lessen their dependence on corporate ride-sharing platforms.[38] They have developed guild-like clubs with rules for membership and conduct, while handling payments through peer-to-peer apps. Workers like these are surviving by governing spaces of their own. But these spaces are improvised and precarious. Policy structures backed by state power, like the one Criscitiello proposes, seem necessary to ensure that workers' self-governance can hold its own against wealthy platform companies.

Governable Infrastructures

The Magnolia Road Internet Cooperative is made up of neighbors who provide internet service for each other, spanning a poorly connected stretch of Rocky Mountain foothills in Colorado.[39] The co-op's closet-sized locker, rented at a local storage facility, holds a supply of routers, wires, and antennas. Consumer-members of the co-op learn to install, use, and repair the equipment themselves. According to the way many people are taught to think about internet access, this does not seem possible—surely such matters are comprehensible only to the national telecom giants that have to be paid to bestow connectivity. But community-governed communications infrastructures, some over a century old, exist throughout rural Colorado and in many parts of the world.

Latin America has a long legacy of *microtelcos*, providing phone and internet service in communities that corporate providers do not see fit to serve.[40] These networks, along with community radio stations, have often been led by women organizing to make their voices heard outside traditional gender roles. The Feminist International Radio Endeavor (FIRE) in Costa Rica, for instance, started with community radio and then went online with the early internet. In a study of Argentinean cases, Paula Serafini argues that community radio stations have served as an ecofeminist "space of care" for communities engaged in ongoing resistance to extractive economies and cultures.[41] For marginalized people, governing infrastructure is itself an act of resistance—in the first place, against others' expectations of what they are capable of.

Feminist scholars have examined how patriarchy mystifies technology, casting it as a domain beyond the possibility of comprehension for all but certain experts. Mystification hides the economics of accumulation that technologies serve, turning people's attention to a marvelous innovation instead of the extraction it enables.[42] As media scholar Lisa Parks has shown, utility firms construct infrastructure so as to be not only incomprehensible but invisible—underground, overhead, or disguised as natural phenomena like trees or rocks. The task of comprehension has required interventions like artist Ingrid Burrington's unofficial guidebook *Networks of New York*, which explains the language and symbols that are intentionally obscure to people who do not work for utility companies.[43] What we cannot understand or notice, we cannot govern.

Infrastructure dictates what people have available to them, on what terms, and at what cost. It requires labor, often shielded from view, to produce and maintain.[44] Corporations do not typically build infrastructure with the intent of enabling users to govern its operation. Yet governable infrastructures have succeeded in addressing the market failures that corporations left in their wake. It was only when the US government began financing electric cooperatives in the 1930s that most rural areas of the country got power lines. Cooperative and municipal broadband systems have advantages of cost and quality over corporate control. Community-based connectivity efforts in many contexts—from Bronx high-rises to towns across Catalonia—continue to show that user-governed networks can succeed where investor ownership falls short.[45]

The dominant allocation of coordination rights, however, often works against governable infrastructure. Many jurisdictions have acceded to corporate-backed laws that prohibit municipal or cooperative broadband services from competing with investor-owned firms. But even where shared ownership is an option, it frequently lacks the access to capital necessary for infrastructure investments. The current pattern of prohibitions could instead be reversed; jurisdictions might prevent long-term investor ownership of physical internet infrastructure. Private firms might build and help capitalize such projects, but the law could ensure that communities of users become the stewards after the build-out. Doing this would require a public commitment to financing access in underserved areas, but as the legacy of rural electrification suggests, such investments are worth the cost.

Software can also serve as infrastructure, particularly the protocols and platforms that large sections of an economy rely on.[46] Governments can support the development of governable platforms by adopting procurement preferences for commons-based software projects, such as the German government's use of Nextcloud for collaborative file-sharing and France's adoption of the Matrix chat protocol. That same software can then be deployed and self-managed by communities anywhere in the world. For instance, an explicitly feminist cooperative in Barcelona, FemProcomuns, uses Nextcloud as part of its Commons Cloud platform.[47] Investments in tools like these enable people to move more of their digital lives into more governable stacks. For both software and hardware infrastructures, public investment can encourage community control over essential services.

Even free and open-source infrastructures, however, can be mystifying in their own right—sometimes even more than proprietary ones. Here, again, feminist tradition calls for a practice of care and pedagogy. Technology education has been a particular focus of Allied Media Projects, the Detroit-based network that formed under the mentorship of Grace Lee Boggs. Instead of the often male-dominated, meritocracy-inclined culture of hackerspaces and hackathons, Allied Media offers the DiscoTech, a model for helping people explore technologies in an intentionally supportive, accessible context.[48] If commons-based infrastructure is expected to work "out of the box" like a commercial product, it may endlessly frustrate us. However, if it comes with a culture of care, it becomes a different kind of tool, one that invites governable spaces.

Self-governance has long been a means of achieving more equitable, accessible infrastructures. But if the value flows of daily online life seem opaque and unknowable, top-down control will seem better than nothing. If governable spaces are a live option, the paternalist promises will reveal themselves for what they are.

PROVISIONING GOVERNABLE SPACE

To lean on self-governance as a policy strategy and to expect people to engage in it means contending with a basic recognition: self-governance takes work.

Perhaps it is asking too much to ask more people to do more of it—especially those who experience marginalization and have tended to receive the brunt of hate speech and abuse online.[49] Must these same people now take on the extra labor of self-governance? Social network CEOs have defended their companies' size and power on the reasoning that scale is necessary to support the costs of protecting users from each other.[50] As the complex of online abuses grows, Silicon Valley leaders—who tend not to hold marginalized identities themselves—insist that they alone can solve these problems of their own making. They are willing to pay for the work. Why shouldn't users accept their offer, however paternalistic?

Feminist economists have long sought to study the present and imagine futures with particular attention to burdens that fall across society unequally. But the feminist response to those burdens has not been to fix them with well-resourced paternalism. Recall, for instance, how the transformative justice activists in chapter 3 respond to harms of policing and incarceration: by taking conflict resolution into their own hands and changing their own communities. Feminist economics likewise tends to begin with the logic of abundance, as opposed to other economists' preoccupation with scarcity.[51] More important than resource limits is the creative potential of people with the resources and support to thrive. People experience self-governance as burdensome when they are not adequately provided for in doing it.

Marilyn Power's summary of feminist economic thought centers on what she calls "social provisioning"—a lens on how social practices organize the distribution of resources and responsibilities.[52] Power identifies five "methodological starting points," which I paraphrase as follows:

- recognizing hidden care and domestic labor in economic life,
- prioritizing human well-being alongside other metrics of wealth,
- correcting for unequal access to authority and agency,
- asserting the validity and inescapability of ethical judgment, and
- intersecting gender analysis with that of race, class, and other forms of identity.

Each of these bears within it a demand for self-governance. Noticing and rewarding invisibilized work or taking participants' ethical reflection seriously— these cannot meaningfully occur without real participant power. Justice in provisioning wealth and in ethical deliberation depends on the presence of governable spaces. Practicing self-governance, once again, depends on having the time, information, and material resources to do so well.

A further aid for rethinking economics beyond paternalism is the legacy of Elinor Ostrom. Although Ostrom did not explicitly identify her work with feminism or employ gender as a guiding concept, she was the first woman to win the Nobel Memorial Prize in Economic Sciences, and her decades-long research on the management of common-pool resources recognized otherwise invisibilized economic practices in which women often play leading roles.[53] Her work may be read as a crusade against illusions of structurelessness, by learning to see and understand how human societies have co-governed land, waterways, and knowledge with well-crafted and recurring structures.[54] She highlights, for instance, the importance of boundary-making, of clear and malleable rules, and of mechanisms for dispute resolution and sanctioning rule-breakers. Online platforms have frequently regarded such practices as cumbersome and antiquated, yet Ostrom's work indicates that they do so at their peril.

Together, Power and Ostrom teach that good governance does not happen by magic or for free. Governable spaces cannot flourish without the leverage to make them meaningful or the resources to make them sustainable. Here, I turn to policy strategies for enlarging the spaces of self-governance in networked life through the mechanisms of territorial government, organizational structures, and the engineering of technologies.

Governments: Ceding Authority

After the Yellow Vest movement swept France in 2018, protesting a regressive fossil fuels tax, President Emmanuel Macron announced a "great debate" in early 2019. It amounted to a nationwide assembly that selected citizens at random to meet together, study economics and climate change, and devise policies to address the environmental crisis more equitably.[55] Participants received payment for their involvement, a kind of provisioning. This was a case of a government seeing fit— or, really, being forced—to provision a governable space.

Juries are familiar today in judicial systems, issuing verdicts within constrained procedures. But not since ancient Athens have they been used widely in the West for legislative or executive functions. Governments have begun to change that, turning to citizen assemblies for solutions to intractable problems ill-suited to partisan legislating. Ireland used an assembly to develop the 2018 referendum that legalized abortion, and Chile convened an assembly to rewrite its constitution entirely; Canada used one to study misinformation online, and an assembly in Michigan developed proposals for addressing COVID-19. Some assemblies are employing complex algorithms to establish a representative selection of members across multiple vectors of identity and experience. In many cases, assemblies have succeeded in cutting through political stalemates after elected representatives failed.[56] But for juries to hold real power presupposes a society in which people really believe that "every cook can govern," to borrow the title of C. L. R. James's essay on jury-based governance. Macron's France was not quite ready for that kind of trust.

As with so many of the assemblies, juries, digital consultations, and community meetings that governments use every day, however, Macron's Citizens Convention for Climate was largely advisory, disconnected from the normal flows of power. As a study of the process put it, "interactions between the citizens [in the assemblies] and the broader public were characterized by mutual scepticism."[57] The process gave Macron an escape hatch from his imperiled climate strategy, and some of its proposals found their way to legislatures. But the process as a whole did not strike most of the French public as legitimate so much as an elite-driven show. To put the matter in terms of social provisioning: the assemblies did not correct the power imbalances that provoked the protests.

Elinor Ostrom stressed that self-governance arises not from abstractions, not from polite consultations but from common resources. Governable spaces must have something at stake. They arise in jurisdictions whose inhabitants share power, where their voices have effect.

It is in principle possible for governments to create governable spaces by carving out domains where direct participation comes with real power, and some do it. This is what happened in 1988, when Porto Alegre, Brazil, invited residents to decide how part of the municipal budget would be spent. This practice of participatory budgeting has since spread to cities worldwide. More recently, after the En Comú coalition gained power in Barcelona in 2015, it introduced a set of participatory processes in keeping with the coalition's social-movement origins. One of the results is Decidim, the open-source software that the city invested in to support its governance experiments—since adopted by other governments around the world. The modules available for Decidim reflect its diverse uses: assemblies, participatory budgeting, structured debates, permissionless initiatives, petitions, juries, delegative voting, crowdsourcing, and more.[58] These are artifacts in code of how people's imaginations move when they have a taste of participatory power.

On the whole, governments have resisted giving up on the pre-digital designs of representative politics. The parties that introduced innovations in Porto Alegre and Barcelona soon lost power, and their governable spaces contracted. Far more often, the door to new forms of effective voice stays closed.

The most promising sites for opening governable spaces may be where the rules have yet to be set, where the necessary jurisdictions do not map neatly to territorial governments. The regulation of online platforms is an example of this, as its challenges transcend localities, and perhaps they require jurisdictions more native to networks. Climate change is another example—a crisis that individual countries have limited incentive to take initiative on but that the human species as a whole urgently needs to confront. Perhaps governments should cede authority over climate governance to a more global jurisdiction, a context where it is less enticing to sacrifice planetary survival for regional benefit. These kinds of issues are frontiers in certain respects, but they need not be subject to feudal homesteading.

In Taiwan, the "digital minister" Audrey Tang has led successful efforts to formulate government policy through digital deliberation, a process she refers to as "listening at scale."[59] Through identifying clusters of public opinion and crowdsourcing proposals with broad support, these efforts seem to bypass the usual partisan talking points and dividing lines. Perhaps this is because Tang's most prominent experiments have dealt with ridesharing apps and COVID-19; apps and viruses that know no borders invite approaches to governance capable of remapping the political terrain.

There are many examples of cross-territorial governance layers already in the making. The municipalist movement is cultivating networks of international cities that have more in common with each other than with their surrounding countrysides. The Global Covenant of Mayors, for instance, enables cities to link their climate commitments independent of national governments. Organizations like the Kurdish Academy of Language connect groups that speak a common language across borders; these may develop methods of shared decision-making and shared standards, just as communities of software developers decide on the features to include in their programming languages. Social-media users, wherever they happen to live, could write the codes of conduct for their platforms—following experiments in crowdsourcing constitutions in places like Mexico City and

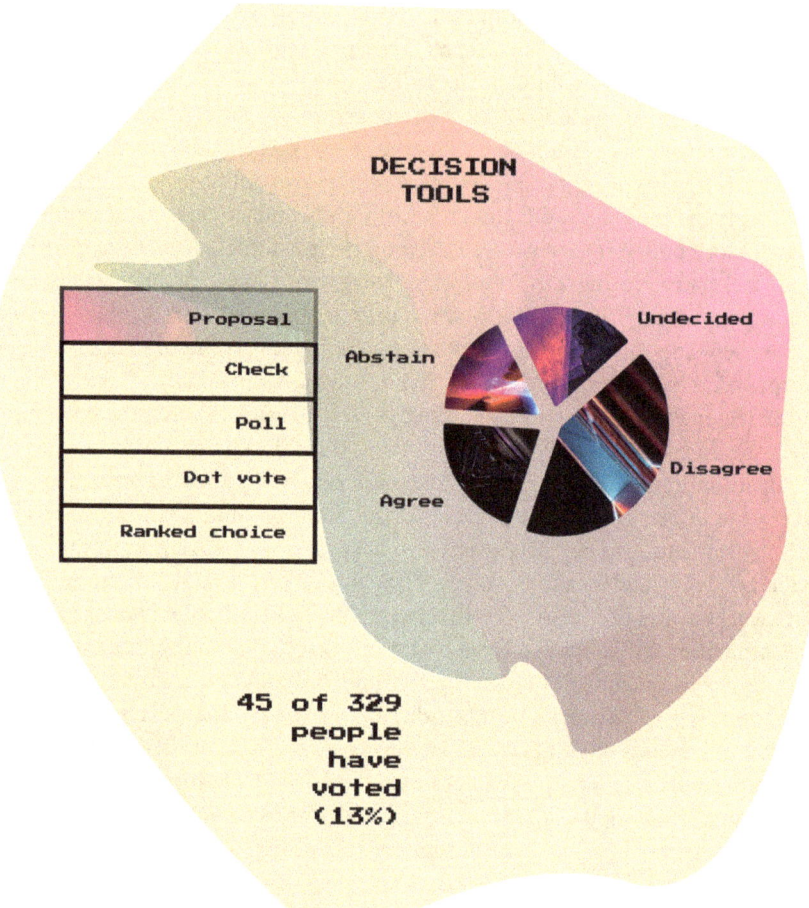

DECISION
TOOLS

| Proposal |
| Check |
| Poll |
| Dot vote |
| Ranked choice |

Abstain
Undecided
Disagree
Agree

45 of 329
people
have
voted
(13%)

FIGURE 14.

Iceland—rather than deferring to the regulations and norms of the country where the platform's servers happen to be.[60]

The advent of blockchains has spurred the plausibility of governments ceding power to governance on networks; by issuing money and enforcing agreements through code, they can do what only governments have been able to do before. Back in 2014, a short-lived startup called Bitnation promised we would all soon have blockchain passports and health insurance. More recently, crypto investor and entrepreneur Balaji Srinivasan published *The Network State: How to Start a New Country*; he envisions token-holders banding together and acquiring land like a corporate retail chain or a religious institution more than a contiguous territory, then securing diplomatic recognition from governments. The mechanism for how power will flow, however, is unclear, and it looks suspiciously like the top-down structure of a Silicon Valley startup.[61] But governments could insist on ceding authority only to network-native polities with strong democratic commitments.

Provisioning new governable spaces begins when existing power structures recognize their own limits, as Macron's government did, and invest some of their powers in other structures that put democracy where it is most needed. This practice is already spreading through the likes of participatory budgeting and citizen assemblies. Governments can similarly organize and cultivate new kinds of spaces for emerging challenges in online life.

Organizations: Delegating Accountability

Online platforms, like governments, have resisted fully provisioning governable spaces. Recall Facebook's act of "democracy theatre" in 2009, when the company held a user referendum on a policy change that was almost surely designed not to reach the quorum that would make it binding.[62] In contrast, when the company now known as Meta formed its Oversight Board a decade later, seeking to deflect ongoing criticisms of its moderation decisions, it did so through an external organization. Although the Oversight Board does not have direct accountability to ordinary users, its rulings create a meaningful check on company behavior. In the future, such an entity might have its members chosen by users, not by the company or its designees.

Another social network under public scrutiny, Twitter, generated a different approach to externalizing power. As at Facebook, the thankless task of moderating content from world leaders and polarizing celebrities had become a liability for the company. In 2019, Twitter established Bluesky, an independent startup devoted to building a decentralized network in which Twitter itself would be only part—handing more possibilities for governance to users, outside the company's reach. Later, after co-founder Jack Dorsey stepped down as Twitter's CEO, he wrote in a text message to future Twitter owner Elon Musk, "A new platform is needed. It can't be a company."[63] This is not a vision that Musk, who renamed the platform X, appears to share, as someone who appears to relish his ability to control the discourse and users' experience at a whim. He discontinued active collaboration with Bluesky, which now operates as a competing app.

Dorsey's perception that there is a mismatch between standard corporate forms and networked life seems to be spreading, even among the most successful beneficiaries of the status quo. In 2018, companies including Uber and Airbnb requested guidance from the US Securities and Exchange Commission for how to distribute company stock to their users in advance of their public stock offerings—particularly the user-workers whom the companies do not regard as employees. The companies offered reasonable explanations for their requests: shared ownership could encourage loyalty and align incentives, just as technology startups habitually offer stock options to early employees. In effect, the regulators said no;

securities law, it seems, knows how to deal with investors and to some degree with employees, but not with users working over networks.[64]

In the following year, 2019, I tried to highlight this challenge by coining the slogan "exit to community," or E2C.[65] Typically, startups backed by venture capital have two options for their inevitable "exit": being acquired by a bigger company or becoming publicly traded on stock markets. Both exit options mean selling to the highest bidder, and any community the startup has built becomes a commodity. The phallic metaphors of "market dominance" and "liquidity event" that populate the jargon of startups guide them toward achieving investor profits more than cultivating healthy communities.[66] E2C is an invitation for startups to explore bringing their most direct participants into structures of ownership and governance. I have worked with dozens of founders attempting to implement it in their companies. They employ mechanisms such as dual-class stock, purpose trusts, cooperatives, nonprofits, and more. We do what we can with what we have. But the E2C meme has spread most widely around blockchains—where conventional securities laws apply ambiguously and where community ownership is, while hardly universal, at least the default setting.

By accompanying startups that want to become governable by their communities, I have seen just how hard this can be under dominant policy regimes. Tokenization through blockchains has been yet another reminder that there can be another way—though blockchains are hardly necessary to achieve shared ownership and governance. Incorporation statutes for companies could be designed to support the flows of shared ownership on cross-border networks, so that users who contribute value can co-own and co-govern the value they create. Financial-system reforms could also enable communities of people with common interests to access capital in ways now available only to companies owned by wealthy investors. If neighbors want to build a broadband network for themselves or if a global network of gig workers wants to own the platform they rely on, they should be able to access financing to do it.[67]

Once again, governable spaces must have social provisioning: the economic and political capacity that self-governance requires. User-governable companies can form with creative entrepreneurship, and they have, but reaching a meaningful share of the larger economy will mean changing the underlying rules.

Some of the largest platforms have already begun to dip their toes into governable waters voluntarily, as in Meta's Oversight Board and the advisory board for hosts that Airbnb created after being unable to issue stock to them directly.[68] These nascent corporate policies remain limited in their power and independence from management, but that could change. Governments might impose a variant of Germany's requirement of worker codetermination through participation on corporate boards;[69] platforms of a certain size might need to have user-elected representatives on their boards or moderation teams. Corporate and securities laws could thereby enshrine governable spaces as a normal aspiration, where the result of successful entrepreneurship is a transition to community control. After a US regulator indicated a standard of being "sufficiently decentralized" for blockchains to operate free of securities regulation, in 2018, crypto projects have had further incentive to distribute ownership and governance widely.[70]

Ownership, like citizenship, is a way of establishing common-pool resources. If organizations are to have owners at all, practicing democracy in them requires democratic ownership. Distributing profits through ownership, also, is a way of provisioning the work of governance, of ensuring that participants can take part because the process is enriching them, not simply draining them. The rights to govern and own should cleave not just to profit-seeking investors but to the users, workers, makers, and lurkers who bring our networks to life.

Technology: Skilling Up

In 2021, a new virtual entity called GitcoinDAO formed to take over control of Gitcoin, a platform that facilitates cryptocurrency donations.[71] Over lunch that summer, founder Kevin Owocki asked me to be a "steward." This meant that I would be included among those to whom DAO token-holders could choose to delegate their voting power. Token-holders included a blend of workers and users who received tokens based on their past contributions and the investors whose capital financed the transition. I turned out to be terrible for the role.

Almost immediately, it was clear that I would fail to keep up with the deluge of information coursing through the DAO's online forum, chat channels, coordination calls, and whisper networks. The only decision I remember voting on was a test poll about pineapple pizza. And yet, in the time since, I have watched as the DAO's ecosystem evolved. A website, daostewards.xyz, provides scorecards on stewards so that token-holders can see how poorly I have been performing. A Steward Council was created to support the most engaged stewards in being more informed and forward-thinking. Interactive primers and informal schools have formed to train new contributors. During times of inflated cryptocurrency markets, the DAO built new software and marketing artifacts for itself furiously; during downturns, it had to make hard decisions and learn discipline, focusing more on the processes among the humans.

Watching GitcoinDAO—just one among thousands of such network-native collectives—is like seeing a new kind of organism searching out its niche. Code and culture are creating each other. The novelty depends on the fact that all this is happening through the power allocated to tokens that can be traded on a distributed network. The jurisdiction of the Ethereum blockchain and the organizational genre of a DAO make possible a more governable stack. Atop those, participants add more layers of software and culture to further hone their self-governance.

This is the kind of cycle that governable spaces can ratchet up: looping back and forth between technological designs and human practices. As the humans develop their political skills, they see new opportunities for software to augment those skills further. Kinds of software appear that would never be built under implicit feudalism—they simply wouldn't be needed or useful. Just as implicit feudalism has built a fortress for itself made of code and norms, governable spaces grow stronger as they reinforce the patterns that make them work. Their mere existence produces demand for more technologies of effective self-governance.

For technology to be governable, users must have the skills to understand its flows of power. Mystification helps keep ostensibly decentralized systems under the control of a small, expert elite. I experienced that in my short career as a steward at Gitcoin, feeling paralyzed in the face of proposals whose context and consequences I didn't understand. Well-intended transparency can mystify when it overloads our attention, when it seems to confuse more than teach. Technologies make policy when their designs dictate what information users do and don't see and how. Technologies make policy in how their interfaces teach us to use them. The skills people need in governable spaces are not simply about how to use the technology, like a user's manual, but about how to craft its policies: what is at stake in the system's design, and what decisions have effects on our lives.

Technology design is policy design. Policies appear in the shape of interfaces, like the steward report cards. But policies also disappear in the underlying infrastructure, in the protocols and incentive structures that lie beneath the surface. Provisioning governable spaces requires not just technologies for governance but also governable technologies, and people equipped to co-design their tools.

FOUNDATIONAL BONDS

In Cadwell Turnbull's story that opened this chapter, there is a theory of change—an explanation for how the residents of St. Thomas became early adopters of Common, the governable computer:

> It shouldn't be surprising that the places most ravaged by climate change are the places where the cooperative commonwealth has been most realized. St. Thomas is one of those places, due in part to the grassroots consensus politics, direct democracy, and cooperative institutions that make up any good solidarity economy, but also plain necessity. Worker cooperatives line St. Thomas' Main Street. Housing cooperatives dot the hillsides of Solberg, Northside, and Bordeaux. Most of the island's grocery stores are multi-stakeholder cooperatives that have strong relationships with local farmers. St. Thomas' many industries are part of regional federations, engaged in worker exchange programs, skill-sharing, and other forms of worker solidarity.[72]

Historically, this is indeed how bursts of cooperative development have tended to go: people conducting local experiments out of necessity band together and build power sufficient to establish public policies, which unlock potential for far more.[73] Prefigure, replicate, and reintegrate into a new normal. Here again are adrienne maree brown's fractals and Alexis de Tocqueville's associations, along with Grace Lee Boggs's belief that activists in the streets of Detroit could save the soul of the country that hollowed out their city. Here again is the modern feminist rediscovery of friendship as a foundational political bond. Democracy starts with seeds, and they grow if we let them.

Against the tenor of most policy discussions, I have insisted throughout this book on the political importance of everyday online life. Attending to everyday life means not ignoring policy but recognizing its connections to our most ordinary encounters. This chapter has stressed that everyday self-governance can be a strategy for policymaking, an approach for confronting many vexing challenges of online life. But doing so requires provisioning: providing resources to support self-governance and ceding power to it.

Turnbull's story is a hopeful one. With the aid of democratic machine-learning, global carbon emissions recede and islanders learn to weather their hurricanes in relative safety. I cannot claim that governable spaces will always turn out so well, at least at first. I offer no such promises. Self-governance is not a solution; it is a practice for problem-solving, and practices can go awry until they find their footing, until their participants learn the skills to manage them. But then, in governable spaces, our difficulties are our own and not someone else's. To have a future more democratic than the present, the structures of power today must embrace that risk.

Epilogue

Metagovernance

The website Native-Land.ca features a map of Earth marked with the territories of Indigenous peoples.[1] The territories appear not as the space between borderlines but as overlapping regions of color. Where I live, three regions intersect, those of the nations known in English as the Arapaho, Cheyenne, and Ute. All three have been stewards of this place. But they have not claimed exclusive domain over it, since their seasonal, migratory ways of life long permitted them to coexist. Speaking from Indigenous North American experience and against today's ascendant nationalism, Glen Coulthard has written, with Matt Hern, of "non-exclusive sovereignties":

> Imagining new renditions of community beyond any transcendent identity is exactly what is required to surpass the brutal nations that stain our times. The idea of "we" can be stripped of its colonial, statist and anthropocentric fixities. It is wholly possible to embrace and refuse identity in the same breath, reaching for a concept of being together that is exposed to the more-than-human. Community needs to extend far past the human if it is to retain any force. It has to think past species and sovereignty as much as flag.[2]

Attempts to imagine new kinds of jurisdictions seem to be on the rise. There is the invisible nation of Wakanda in the *Black Panther* universe and Janelle Monáe's android city Metropolis. Online communities give themselves jurisdictional names that refer to old kingdoms and regions of outer space.[3] In an age beset with storms, fires, and extinctions signaling ecological breakdown, which existing regimes have been unwilling to reverse, the longing for other jurisdictions comes easily. The evidence of governance failures is everywhere, and our imaginations need somewhere else to go.

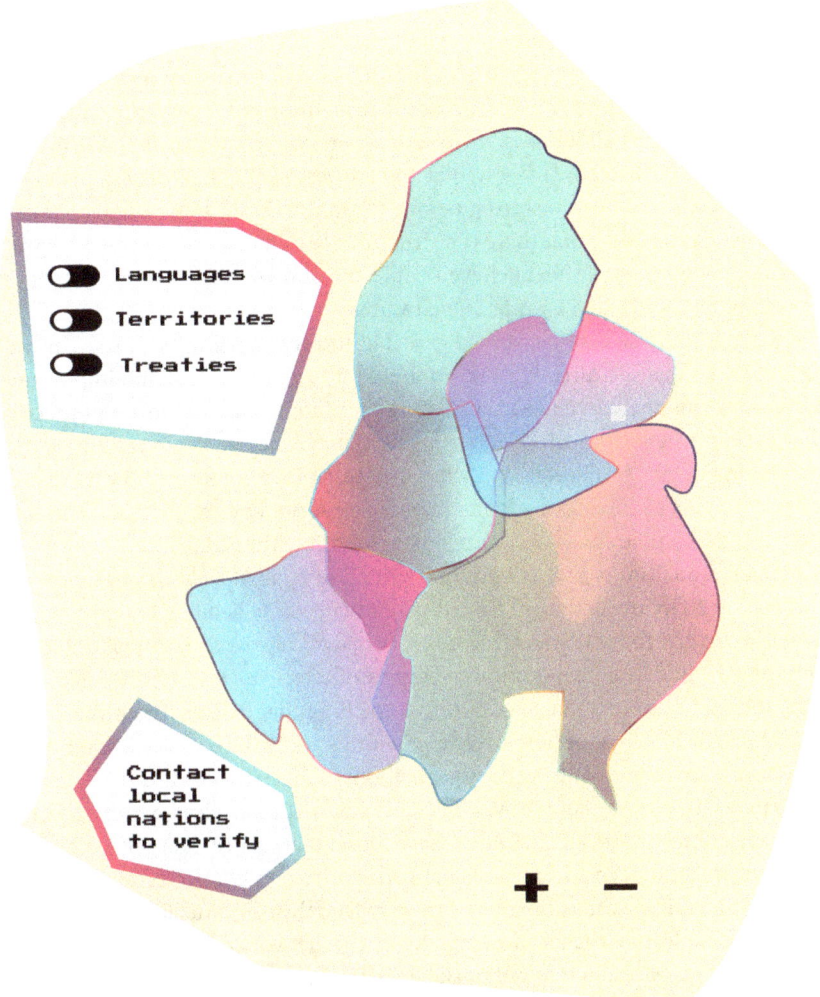

Languages

Territories

Treaties

Contact
local
nations
to verify

\+ −

FIGURE 15.

One way to understand the global turn to ethnonationalist fantasies and strongman-style leaders is as a symptom of the nation-state's weakness. If governments continually fail to deliver the governance we actually need, we can at least feel better by doubling down on what governments are capable of still. Nation-states can't stop climate change, but they can enact a nostalgic fantasy of national identity and militarize their borders. Politicians can impress their constituents by denouncing transgender kids and insulting racialized minorities. They can invade less powerful countries for inexplicable reasons. But other paths are possible. Instead, the nation-state could offload some of its governance burdens to new kinds of jurisdictions, concurrent layers of governance that are more tailored and accountable in their domains.

In an essay on the idea of post-nationalism, adrienne maree brown describes a sensation of liberation in trying to imagine something other than the usual tools of governance—the ballot boxes, the political parties—toward tools closer to what communities really need: "i know that the hardest step is not getting people to choose the best tools, but inspiring people to want to build something at all. and then, growing the belief that there is a structure they could cocreate in which they could belong without battle. i believe people can and will demand better tools as they fall in love with their own possible futures."[4]

Governable spaces are steps into possible futures, starting with the connective networks that are already now among us. I have argued for the importance of everyday online experience as a starting point for imagining through practice what it might be like to more fully, appropriately, deeply co-govern the world. The making of governable spaces invites questions I do not know how to answer. But this book is an attempt to frame the conversation, first by going back to how governance in online space has been for too long constrained.

Implicit feudalism is a design pattern written in code, upheld through network protocols, the circumvention of labor law, and corporate liability. It shapes and is shaped by, in turn, the practices of billions of people taking part in online social spaces every day. Feudal designs permit expressions of affective voice but not, for most users, the more direct leverage of effective voice. Feudal defaults teach their users the embedded ideology of homesteading, with its roots in the colonization of the American West, now colonizing the imaginations of the networked world. The result has been not the democratic revival that early internet evangelists promised but an authoritarian turn. Demagogic leaders have found that, better than pro-democracy protesters, they can take advantage of network norms and flows to seize power over societies increasingly accustomed to everyday feudalism. Homeplaces arise everywhere, as people rub against the defaults of the networks to create accountable and nourishing spaces. But do not mistake the resistance for the regime.

Some kinds of resistance set out to replace the regime, to unravel the habits of feudalism with self-governance. Through governable stacks, communities can identify and root out feudal patterns and remake them as commons. The stack is social, technical, and environmental infrastructure. It is affective and effective. It is economic and spiritual. Through relations of subsidiarity, communities can work out their stacks as they see fit locally, while still participating in larger networks. Through modular designs, they can copy and adapt practices from elsewhere and share their creations with others. Through an archaeology of past governance practices, infused with a spirit of ancestry, the wideness of possibility grows beyond tech culture's relentless and self-limiting fascination with innovation.

None of this can get very far without policies in the background that support community ownership, that respect self-governance over paternalistic decrees.

Even well-meaning policy can maintain the mystification of technology and fail to cultivate the political skills of people. Political skills can form only if we practice them in our most ordinary interactions. The everyday and the institutional reflect back on each other like a fractal house of mirrors. One does not begin without the other, and to that extent I cannot claim to offer an orderly program so much as a bidding to try everything at once.

Permit me one last concept. Political theorists began writing about *metagovernance* in the 1990s—as Bob Jessop defines it, "the organization of the conditions for governance in its broadest sense," or simply the "organization of self-organization"[5] Addressing failures of governance, this coterie recognized, requires more than simply considering a certain situation in isolation, because there is a broader context in which it occurs. The term arose as scholars began to see governance proliferating across domains more widely than they had noticed before, spreading across public and private institutions, especially through schemes for economic development. They noticed that international norms and rules were orchestrating and constraining the range of possibilities.

Implicit feudalism has been a kind of metagovernance, and like so much metagovernance, it too often hides beneath our notice. But changing how we govern requires being attentive to the metagovernance at work. Shaping the background conditions of governance is itself a form of governance.

Independently, people in crypto have begun talking about metagovernance, too—less as a theory than as daily practice. For them, metagovernance happens when one DAO, for instance, holds tokens from other DAOs. That means a proposal vote in that first DAO might cascade to other DAOs, across the ecosystem and potentially back again.[6] DAOs habitually engage in token swaps to solidify collaborations, exchanging power in each other. Products like dashboards and voting tools are appearing specifically to support the resulting kinds of many-sided governance. In these contexts, metagovernance means trying to comprehend a condition of bewildering integration.

What I have been up to all along in this book is a kind of metagovernance, a critique of feudal habits and a call for cultivating democratic communities. I have argued that the design of metagovernance for online spaces matters immensely. Social networks so far have fed democratic erosion and an authoritarian turn, but other kinds of design could draw people toward a world of non-exclusive sovereignties, as Coulthard and Hern put it, and of right relation to the more-than-human. More intentional metagovernance can veer human societies toward accountable connectivity and toward the planet now asking us to get our act together, or else.

What kinds of interfaces, power structures, and skills will guide us in a world where we can co-govern more and more of the jurisdictions we inhabit? A different sort of design, a different tenor of education, and different practices of attention will all be necessary. The Chilean president Salvador Allende attempted to make his country a governable space with Project Cybersyn, a pre-internet computer network headquartered at a stylized nerve center in the capital.[7] How would the world look and feel if each of our communities were a Cybersyn, a convergence of transparent information flows and decisions, under democratic control?

As Native-Land.ca suggests, a habitable vision for the future can begin with recognizing more fully the legacies we stand among. *Metagovernance* is academic jargon and memetic vernacular all at once, and from both directions the word gestures toward a struggle for democracy many of us have not yet noticed is happening. What is at stake? Something no bigger than we are. Governable spaces are a starting point for becoming, together, more fully ourselves.

ACKNOWLEDGMENTS

The making of this book has been far from a democracy, but many hands have shaped it nonetheless.

My daily teachers in the craft of self-governance are my family, especially Claire, Daniel, and Sylvia, and before that Barbara and Mitchell. I dedicate the book to my grand-father, Alan Schneider, who came to the end his life while I was beginning to write. I hope to be for my students someday who he was for his and to have such trust as he did in the young.

The Department of Media Studies at the University of Colorado Boulder has been an-other governable space in my daily life—a community of mentors and comrades, group chats and bowling matches that carries on the precious inheritance of the academic guild with whatever tools we can find. My students, particularly those in the Media and Public Engagement MA program and the Media Economies Design Lab, keep the questions at work in this book alive in me through their restlessness and resolve. The Center for Media, Religion, and Culture enables me to be a student again during our seminar each week.

It has been a pleasure crafting the illustrations with Darija Medić, a brilliant artist and scholar. We made our breakthroughs while my kids played in the game room of the Media Archaeology Lab on our campus, during lulls in her Sunday afternoon shifts.

As I wrote this book, the Metagovernance Project grew from five of us writing a paper to upwards of a thousand people sharing an online space. That community has been a source of constant inspiration and correction, software projects and house parties, schemes and swag. I can no longer keep track of all that we are up to together. I hope this book helps explain to the wider world why what we all care about matters.

I am grateful to publish again with University of California Press, which took a chance on me many years ago with my first two books. Thank you to Michelle Lipinski and LeKeisha Hughes for shepherding this project there and to Enrique Ochoa-Kaup for tak-ing an initial interest. Barbara Armentrout was a rigorous and thoughtful copyeditor. My former student, Laura Daley, produced the index. My agent, David Patterson, guided me through the process once again with wisdom and deliberateness in even the most seemingly innocuous of phrases.

This book benefited immensely from its patient early readers. I received valuable feedback and encouragement on drafts from academic colleagues, including Nabil Echchaibi, Zizi Papacharissi, Erica Robles-Anderson, Adrienne Russell, Ted Striphas, and, with particular generosity and gusto, Ethan Zuckerman. I am grateful as well to the numerous readers and reviewers of the previously published articles that fed the book, listed below; readers are credited where those are published. I also had the benefit of comments from my students Hari Anantharaman and Nikita Menon. Josh Dávila provided helpful feedback on the crypto material.

The Metagovernance Project hosted a collective review process for a draft of the book, thanks to the skillful facilitation of Cent Hosten. Reviewers included Hosten as well as Seth Frey, Monika Jankowska, Lucia Korpas, Tara Merk, Kelsie Nabben, Michael Holton Price, Ellie Rennie, Philip Sheldrake, Joshua Tan, Aashka Tank, Anna Weichselbraun, Amelia Winger-Bearskin, Michael Zargham, and Jacky Zhao. Shauna Gordon-McKeon provided especially extensive, clarifying feedback to which I hope my revisions have done some justice.

The backbone of this book is a set of articles that I have woven together into an extended argument. The book includes portions of them, which have been edited and reworked. I am grateful to everyone who reviewed and edited those articles, and contributors are listed where they are published.

- Chapter 1: "Admins, Mods, and Benevolent Dictators for Life: The Implicit Feudalism of Online Communities," *New Media & Society* 24, no. 9 (2022), https://doi.org /10.1177/1461444820986553.
- Chapter 2: "Homesteading on a Superhighway: The Californian Ideology and Everyday Politics," *International Journal of Communication* 17 (2023), https://ijoc.org /index.php/ijoc/article/view/19342.
- Chapter 3: "Is Democracy Sacred? Case Studies in Political Imagination," in *Hypermediations: Religion, Media, and Crisis*, edited by Nabil Echchaibi, Stewart Hoover, Nathan Schneider, and Deborah Whitehead (forthcoming); "Cryptoeconomics as a Limitation on Governance," Mirror (August 11, 2022), https://ntnsndr .mirror.xyz/zO27EOn9P_62jVlautpZD5hHB7ycf3Cfc2N6byz6DOk; "Bits of Moloch," *Machines in Between* podcast (November 19, 2022), https://www.machinesinbetween .com/segments/bits-of-moloch.
- Chapter 4: "Governable Stacks against Digital Colonialism," *TripleC: Communication, Capitalism & Critique* 20, no. 1 (January 12, 2022), https://doi.org/10.31269/triplec .v20i1.1281.
- Chapter 5 and epilogue: "Governable Spaces: A Feminist Agenda for Platform Policy," *Internet Policy Review* 11, no. 1 (March 22, 2022), https://doi.org/10.14763 /2022.1.1628; "The Tyranny of Openness: What Happened to Peer Production?" *Feminist Media Studies* 22, no. 6 (2021), https://doi.org/10.1080/14680777.2021.1890183; "Lighten the Load of the Nation-State," *Kernel* no. 2 (2022), https://www.kernelmag .io/2/lighten-nation-state.

I also drew heavily on but did not directly include material from these coauthored articles: Seth Frey and Nathan Schneider, "Effective Voice: Beyond Exit and Affect in Online Communities," *New Media & Society* (2021), https://doi.org/10.1177/14614448211044025; Amy Hasinoff and Nathan Schneider, "From Scalability to Subsidiarity in Addressing Online Harm," *Social Media + Society* 8, no. 3 (2022), https://doi.org/10.1177/20563051221126041; Nathan Schneider et al., "Modular Politics: Toward a Governance Layer for Online Communities," *Proceedings of the ACM on Human-Computer Interaction* (April 2021), https://doi.org/10.1145/3449090. Again, each of those articles acknowledges feedback on early drafts where they are published.

The project profiles in this book each reflect complex collaborations; none of them would have been possible if it had been me alone.

- CommunityRule: Drew Hornbein contributed immensely to the visual and software design, as well as designing the accompanying book, along with further contributions by Asher Farr and Deacon Rodda. A grant from the Smart Contract Research Forum supported a key phase of development.
- A People's History of Twitter: This emerged through my longtime collaboration with Danny Spitzberg, who managed the project with support from a large team, including Mara Abrams, Jacky Alciné, and Wagatwe Wanjuki. We are grateful to all the organizations that contributed to funding the effort.
- Excavations: This residency was a collaboration with Federica Carugati and Darija Medić. It was possible through a grant from the Eutopia Foundation.
- Modpol: Luke Miller was my design partner and did major development work, with documentation support by Skylar Hew. Development began with the support of a residency at the Bentway Conservancy.

This book includes material based on work supported by the National Science Foundation under grant no. 2217654. Any opinions, findings and conclusions, or recommendations expressed in this material are mine and do not necessarily reflect the views of the NSF.

Finally, I want to acknowledge the Open Monograph Ecosystem Initiative at the University of Colorado Boulder Libraries, which provided the financial support to make this book available open-access through the University of California Press's Luminos program. It is a privilege to write in the context of colleagues working to create the free, accessible information commons that our governable spaces need.

NOTES ON ILLUSTRATIONS

The illustrations in this book, created by Darija Medić, evoke the online spaces discussed therein, often combining elements from various interfaces. The font for text on the illustrations is PressStart by Cody Boisclair. The following sources were used to inform each illustration:

INTRODUCTION: DEMOCRACY IN THE WILD

1. This is a question that evokes, for instance, Jay-Z and West, "No Church in the Wild," along with Norton, *Wild Democracy*, a book that resonates with much of my thinking throughout these pages. I hope to echo Norton's insistence on the courage, risk, rebellion, and creativity that vibrant democracy requires.

2. The naive simplicity of my working definition for *governance* is not to take away from the significant scholarship around the concept and terminology, such as Bevir, *Key Concepts in Governance*; Levi-Faur, *The Oxford Handbook of Governance*. That literature engages largely in the sphere of political science and institutional theory, however, and its high degree of specificity is less useful in the comparatively underdeveloped realm of online user-governance that is the primary focus here. For now, indeterminacy helps preserve a strategic breadth to a discussion that I argue is only beginning.

3. My understanding of democracy as a horizon is indebted to Jacques Derrida's sense of democracy "to come"; see Dinan, "Keeping the Old Name."

4. Alexander et al., "Online Exchange on 'Democratic Deconsolidation'"; Diamond, "Democracy's Arc"; Foa and Mounk, "The Signs of Deconsolidation"; Papacharissi, *After Democracy*; Silva-Leander, *Global State of Democracy Report 2021*; Wike, Silver, and Castillo, *Many People around the World Are Unhappy with How Democracy Is Working*. See also democratic-erosion.com, a multi-university collaborative course on the topic.

5. V-Dem Institute, *Democracy Report 2022*, 6.

6. Knight, Schor, and Jorgenson, "Wealth Inequality and Carbon Emissions in High-Income Countries"; Carayannis, Campbell, and Grigoroudis, "Democracy and the Environment."

7. Andrias, "Separations of Wealth"; Bartels, *Democracy Erodes from the Top*; Chi and Kwon, "The Trust-Eroding Effect of Perceived Inequality."

8. Applebaum and Pomerantsev, "How to Put Out Democracy's Dumpster Fire"; Gurri, *The Revolt of the Public and the Crisis of Authority in the New Millenium*; Haidt, "Why the Past 10 Years of American Life Have Been Uniquely Stupid"; Lorenz-Spreen et al., "A Systematic Review of Worldwide Causal and Correlational Evidence on Digital Media and Democracy"; Persily and Tucker, *Social Media and Democracy*; Sunstein, *#Republic*; Törnberg, "How Digital Media Drive Affective Polarization through Partisan Sorting"; Vaidhyanathan, *Antisocial Media*. See also this unpublished, collaborative bibliography evaluating evidence on the topic: Haidt and Ball, "Social Media and Political Dysfunction."

9. Alkhatib, "To Live in Their Utopia"; Gillespie, "Content Moderation, AI, and the Question of Scale"; Katsaros et al., "Procedural Justice and Self-Governance on Twitter"; Myers West, "Censored, Suspended, Shadowbanned"; Nurik, "'Men Are Scum'"; Savolainen, "The Shadow Banning Controversy."

10. On public regulation, see, for example, Balkin, "How to Regulate (and Not Regulate) Social Media"; Boxell and Steinert-Threlkeld, "Taxing Dissent"; Casarosa, "Transnational Collective Actions for Cross-Border Data Protection Violations"; Chu, "Censorship or Protectionism? Reassessing China's Regulation of Internet Industry"; Rochefort, "Regulating Social Media Platforms." On corporate regulation, see, for example, Gillespie, *Custodians of the Internet*; Klonick, "The New Governors"; Seering, "Reconsidering Self-Moderation."

11. Papacharissi, *Affective Publics*; Schneider, *Thank You, Anarchy*; Sitrin, Azzellini, and Harvey, *They Can't Represent Us!*; Tufekci, *Twitter and Tear Gas*.

12. All subsequent quotations from Tocqueville, *Democracy in America*, bk. 2, sec. 2, ch. 7.

13. Matsusaka, *Let the People Rule*; Putnam, Leonardi, and Nanetti, *Making Democracy Work*; Putnam and Garrett, *The Upswing*; Watkin, Gerrand, and Conway, "Introduction: Exploring Societal Resilience"; Zuckerman, "How Social Media Could Teach Us to Be Better Citizens."

14. Does and Jacquet, "Small-Scale Deliberation and Mass Democracy"; Johnson, Carlson, and Reynolds, "Testing the Participation Hypothesis"; Niemeyer, "Scaling Up Deliberation to Mass Publics"; Pateman, *Participation and Democratic Theory*; Wu and Paluck, "Participatory Practices at Work Change Attitudes and Behavior toward Societal Authority and Justice."

15. Gordon Nembhard, *Collective Courage*; Goodwyn, *The Populist Moment*; Holyoake, *The History of Co-Operation*; Thompson, "Frederick Douglass and Co-ops in 1846." W. E. B. Du Bois also studied and advocated cooperative enterprise as part of his vision for "abolition democracy" (discussed in chapter 3). See, for instance, Du Bois, *Black Reconstruction in America*, chs. 9 and 14.

16. Fung and Wright, *Deepening Democracy*; Wright, *Envisioning Real Utopias*, ch. 10. Shortly before his death, I had the opportunity to attend a meeting where Wright exchanged his ideas on "real utopias" with cooperativists in Italy: "Pathways to a Cooperative Market Economy" at the University of Padua in 2017.

17. Felski, "The Invention of Everyday Life"; Highmore, *Everyday Life and Cultural Theory*; Tria Kerkvliet, "Everyday Politics in Peasant Societies (and Ours)"; and most significantly for my purposes, Agre, "The Structures of Everyday Life." This discussion continues in greater detail in chapter 2.

18. Norton, *Wild Democracy*, 37. She holds this notion of the commonplace in opposition to Carl Schmidt's influential identification of sovereignty with the "exception."

19. Atanassow, "Colonization and Democracy"; Welch, "Colonial Violence and the Rhetoric of Evasion."

20. James, *The Black Jacobins*; James, "Every Cook Can Govern."

21. Boggs, *Living for Change*; Boggs and Kurashige, *The Next American Revolution*; King, "Introduction to Boggs."

22. brown, *Emergent Strategy*; brown, *Holding Change*; brown, *Pleasure Activism*.

23. Quijano, "Coloniality and Modernity/Rationality," 177.

24. I refer to *feedback loops* in the sense of systems theorists such as Meadows, "Leverage Points," as well as the cybernetics tradition discussed later in this introduction. For an influential work on the inspirational qualities of fungi, see, for instance Sheldrake, *Entangled Life*.

25. Winner, "Do Artifacts Have Politics?"; Wyatt, "Technological Determinism Is Dead; Long Live Technological Determinism."

26. Gillespie, *Custodians of the Internet*, 212.

27. Agre, "The Practical Republic" (discussed at much greater length in chapter 2); Albergotti, "He Predicted the Dark Side of the Internet 30 Years Ago"; Benjamin, *Race after Technology*, 17, where the phrase appears in italics; it stands in contrast to Facebook's slogan "Move fast and break things." See also Schneider, "The Joy of Slow Computing."

28. Hepp, *Deep Mediatization*, 5. See also Couldry and Hepp, *The Mediated Construction of Reality*.

29. Hepp, *Deep Mediatization*, 105, 109. The use of *recursion* comes from Kelty, *Two Bits*.

30. Kember and Zylinska, *Life after New Media*, xiii, xv, xvii, and the book's conclusion, "Creative Media Manifesto." I draw on their work also in Schneider, "Mediated Ownership," as the basis for redirecting the mediation of economic capital, reconceptualized as a form of media.

31. Kirby and Emerson, "As If, or, Using Media Archaeology to Reimagine Past, Present, and Future"; Wershler, Emerson, and Parikka, *The Lab Book*. I am grateful for the hospitality of the lab's founding director Lori Emerson and the wizardry of managing director libi rose striegl. For media archaeology more generally, see Huhtamo and Parikka, *Media Archaeology*.

32. Papacharissi, *After Democracy*, ch. 1.

33. Papacharissi, *After Democracy*, ch. 5. See also Gargarella, "From 'Democratic Erosion' to 'a Conversation among Equals.'"

34. Winner, "Do Artifacts Have Politics?," 135.

35. Cherry, "Beyond Misclassification"; Scarry, *Thermonuclear Monarchy*.

36. Important reference points on democratic innovation in governments include Barandiaran et al., "Decidim"; Bernal, "How Constitutional Crowdsourcing Can Enhance Legitimacy in Constitution Making"; Bria, "Barcelona Digital City"; Chwalisz, *Innovative Citizen Participation and New Democratic Institutions*; Chwalisz, *The People's Verdict*; Dixon and Piepzna-Samarasinha, *Beyond Survival*; Fung and Wright, *Deepening Democracy*; Giraudet et al., "'Co-construction' in Deliberative Democracy"; He and Warren, "Authoritarian Deliberation"; Hsiao et al., "vTaiwan"; Landemore, *Open Democracy*; Pogrebinschi, *Innovating Democracy?*; Reybrouck, *Against Elections*; Sousa Santos, "Participatory Budgeting in Porto Alegre"; Stempeck, "Next-Generation Engagement Platforms"; Tsai, *Accountability without Democracy*; Tseng, "Algorithmic Empowerment"; Whittington, *Democratic Innovation and Digital Participation*; Youngs and Godfrey, "Democratic Innovations from around the World."

37. Important reference points on democratic innovation in crypto include Alston, "Constitutions and Blockchains"; Buterin, *Proof of Stake*; Catlow and Rafferty, *Radical Friends*; Komporozos-Athanasiou, *Speculative Communities*; Reijers et al., "Now the Code Runs Itself"; Swartz, "Blockchain Dreams"; Zargham and Nabben, "Aligning 'Decentralized Autonomous Organization' to Precedents in Cybernetics." By way of disclosure: I hold meaningful amounts of cryptocurrency, largely due to a modest contribution to the initial Ethereum crowdsale, which I was participant-observing at the time as a reporter.

38. Escobar, *Designs for the Pluriverse*; Costanza-Chock, *Design Justice*. See also the work of the Verses collective (verses.xyz), which has introduced language of "pluriverse" to crypto culture, and the Design Justice Network (designjustice.org), a community of design practitioners.

39. Allen, *Tocqueville, Covenant, and the Democratic Revolution*; Bollier and Helfrich, *Patterns of Commoning*; Börgers, *An Introduction to the Theory of Mechanism Design*; Ostrom, *Governing the Commons*; Swann, *Anarchist Cybernetics*. Ostrom initially described her framework as "Institutional Analysis and Design" before changing *Design* to the more inclusive *Development*. *Cybernetics*, in turn, derives from the Greek for "governance."

40. Illich, *Tools for Conviviality*, 11, 10.

41. See metagov.org, and our initial research paper, Schneider et al., "Modular Politics."

42. The Catholic Worker quote is a saying of the movement's co-founder Peter Maurin, as quoted in Day, *The Long Loneliness*, 170. On the relevance of Catholic social teaching to democratic theory and practice, see Bretherton, "Democracy, Society and Truth"; Schneider, "'Truly, Much Can Be Done.'" On the cooperative movement, see Schneider, *Everything for Everyone*.

43. See, for instance, Brennan, *Against Democracy*; Parvin, "Democracy without Participation."

1. IMPLICIT FEUDALISM: THE ORIGINS OF COUNTER-DEMOCRATIC DESIGN

1. #BLM10, "It Is Time for Accountability."

2. Zuckerberg, "Mark Zuckerberg's Letter to Investors." On the company's share structure, see Han, "The Facebook IPO's Face-off with Dual Class Stock Structure."

3. Norton, *Wild Democracy*, 141–45, reflects on both the way in which "feudalism" was named as a period only after it was supposedly over so as to put it in the past, even while its remnants remain with us: "Perhaps feudalism has been broken, but if so, we still walk among its shards" (142). Also relevant, far beyond the digital context, is Graeber and Sahlins, *On Kings*, which probes the apparent ubiquity of kingship—and the longing for simple, hierarchical power structures—as a cosmopolitical arrangement.

4. Stasavage, *The Decline and Rise of Democracy*, is a history of premodern democracy across a wide variety of societies and democratic forms.

5. Shaw and Hill, "Laboratories of Oligarchy?," observes this "iron law" at work among communities on the commercial wiki platform Wikia. Frey and Sumner, "Emergence of Integrated Institutions in a Large Population of Self-Governing Communities," finds similar practices in a large study of Minecraft servers.

6. Huhtamo and Parikka, *Media Archaeology*, 3. See also Parikka, *What Is Media Archaeology?*; Malloy, "The Origins of Social Media."

7. Ronzhyn, Cardenal, and Batlle Rubio, "Defining Affordances in Social Media Research," 14. See also Nagy and Neff, "Imagined Affordance"; Evans et al., "Explicating Affordances."

8. On disaffordances (and dysaffordances), see Costanza-Chock, *Design Justice*, ch. 1; Wittkower, "Principles of Anti-discriminatory Design."

9. Fansher, Chivukula, and Gray, "#Darkpatterns"; Mathur, Kshirsagar, and Mayer, "What Makes a Dark Pattern . . . Dark?"

10. Hirschman, *Exit, Voice, and Loyalty*.

11. Matias, "Quitting Facebook & Google," documents Janet Vertesi's harrowing attempt to avoid surveillance-based technology.

12. Frey and Schneider, "Effective Voice."

13. Christensen and Seuss, "Hobbyist Computerized Bulletin Board"; Rheingold, *The Virtual Community*; Malloy, "The Origins of Social Media."

14. *BBS: The Documentary*, episode 2, 11:30 and 17:20.

15. *BBS: The Documentary*, episode 2, 6:35 and 16:40.

16. Quotations from Driscoll, "Thou Shalt Love Thy BBS." See also Driscoll, *The Modem World*, particularly for a fuller account of the governance diversity and impact of the BBS era. While I center the role of BBS communities in shaping future norms, one could also consider narratives that center academic networks (e.g., Wooley, "PLATO: The Emergence of Online Community") and political economy (e.g., O'Mara, *The Code*). I believe BBS practices were particularly important in shaping the hobbyist networks that would have been the most likely site for democratic structures to emerge.

17. Christensen and Seuss, "Hobbyist Computerized Bulletin Board."

18. See Castillo, "VOTEMGR"; OneNet, "The OneNet Member Constitution."

19. Wade, *The Anarchist's Guide to the BBS*, 69. See also Ciarcia, "Turnkey Bulletin-Board System," 97.

20. Dibbell, "A Rape in Cyberspace."

21. Ludlow, *Crypto Anarchy, Cyberstates, and Pirate Utopias*, 324. Thanks to Amy Bruckman for sharing her experience with me from Xerox PARC; she pointed out that "wizardly fiat" was rarely used, and most often because features requested through user governance were outside the capacity of the maintainers to implement.

22. Malloy, "The Origins of Social Media"; Pfaffenberger, "'If I Want It, It's OK'"; Spencer and Lawrence, *Managing Usenet*.

23. Pfaffenberger, "'If I Want It, It's OK,'" 376. See also Big-8.org, "Big-8 Usenet Hierarchies."

24. Big-8.org, "Moderated Newsgroups."

25. Kollock and Smith, "Managing the Virtual Commons."

26. Pfaffenberger, "'If I Want It, It's OK,'" 374–75. Pfaffenberger also concludes with an assessment of how technological designs powerfully shaped the possibilities of human politics on the system.

27. Hyman, "Twenty Years of ListServ as an Academic Tool." On ARPANET and email more generally, see Hafner and Lyon, *Where Wizards Stay Up Late*.

28. Caines, "How to Run a Mailing List." Brackets in the original.

29. Wooley, "PLATO: The Emergence of Online Community."

30. Reid, "Communication and Community on Internet Relay Chat"; Rintel and Pittam, "Strangers in a Strange Land."

31. McPherson, "US Operating Systems at Mid-Century."

32. Conger, "'Master,' 'Slave' and the Fight over Offensive Terms in Computing"; Eglash, "Broken Metaphor." The 2020 wave of the Black Lives Matter movement prompted a widespread turn away from using such racialized terms in software systems.

33. Benkler, *The Wealth of Networks*.

34. Federman, "The Penguinist Discourse"; Laffan, "A New Way of Measuring Openness."

35. Jones, "Linux Founder Takes Some Time Off to Learn How to Stop Being an Asshole."

36. Tourani, Adams, and Serebrenik, "Code of Conduct in Open Source Projects."

37. Ehmke, "Codes of Conduct."

38. Fish et al., "Birds of the Internet"; Coleman, *Coding Freedom*. Although protocol governance is outside the scope of this narrative, consider also the practice of "rough consensus" at the Internet Engineering Task Force; see Resnick, "On Consensus and Humming in the IETF."

39. Cosentino, Izquierdo, and Cabot, "Three Metrics to Explore the Openness of GitHub Projects."

40. Apache Software Foundation, "Incubating Issues."

41. Reagle, *Good Faith Collaboration*, ch. 6.

42. Kostakis, "Peer Governance and Wikipedia"; Jemielniak, *Common Knowledge?*

43. Wikipedia, "Wikipedia: Role of Jimmy Wales."

44. Shaw and Hill, "Laboratories of Oligarchy?"

45. Halfaker et al., "The Rise and Decline of an Open Collaboration System."

46. Freeman, "The Tyranny of Structurelessness." On the essay's afterlife in tech culture, see Cohen, "A 1970s Essay Predicted Silicon Valley's High-Minded Tyranny."

47. Tufekci, *Twitter and Tear Gas*, 272.

48. Zacchiroli, "Debian."

49. Double Union, "Base Assumptions." When I visited Double Union in the mid-2010s, a similar phrase served as the Wi-Fi password.

50. Alba, "Mark Zuckerberg Is Sure Acting Like Someone Who Might Run for President"; Zuckerberg, "Building Global Community."

51. Gillespie, "The Politics of 'Platforms.'"

52. Kosseff, *The Twenty-Six Words That Created the Internet*.

53. Coate, "Cyberspace Innkeeping"; Seering et al., "Moderator Engagement and Community Development in the Age of Algorithms."

54. Margonelli, "Inside AOL's 'Cyber-Sweatshop'"; Terranova, "Free Labor."

55. Wang et al., "Governing for Free," presents a nuanced exploration of how different governance regimes generate motivating psychological effects for moderators, based on a survey of Reddit users. Top-down governance models provide some kinds of psychological benefits, particularly in large communities, while more democratic models can provide others.

56. Lampe and Resnick, "Slash(dot) and Burn"; Ganley and Lampe, "The Ties That Bind."

57. Seering et al., "Moderator Engagement and Community Development in the Age of Algorithms"; Leavitt and Robinson, "The Role of Information Visibility in Network Gatekeeping."

58. Jhaver et al., "Human-Machine Collaboration for Content Regulation."

59. On Zuckerberg and corporate governance, see Stewart, "Facebook Will Never Strip Away Mark Zuckerberg's Power." On "democracy theatre," see Bonneau et al., "Democracy Theatre."

60. Matias, "Going Dark"; Massanari, "#Gamergate and The Fappening"; Fiesler et al., "Reddit Rules!"

61. Conway, "How Do Committees Invent?" See also the formalization of the concept as the "mirroring hypothesis" in Colfer and Baldwin, "The Mirroring Hypothesis."

62. Zuckerberg, "A Privacy-Focused Vision for Social Networking."

63. Newport, "TikTok and the Fall of the Social-Media Giants."

64. Kelty, *The Participant*, 1.

65. On subreddit distribution, see Medvedev, Lambiotte, and Delvenne, "The Anatomy of Reddit." On Minecraft servers, see Frey and Sumner, "Emergence of Integrated Institutions in a Large Population of Self-Governing Communities."

66. Katsaros et al., "Procedural Justice and Self-Governance on Twitter"; Myers West, "Censored, Suspended, Shadowbanned"; Nurik, "'Men Are Scum.'"

67. For a thorough discussion of institutional diversity as a principle in governance design, see Ostrom, *Understanding Institutional Diversity*.

68. Hirschman, *Exit, Voice, and Loyalty*; McEwan and Gutwin, "A Case Study of How a Reduction in Explicit Leadership Changed an Online Game Community." See also Seering, "Reconsidering Self-Moderation," on the opportunity for more research on community-based moderation.

69. Fritz, *The Theory of the Mixed Constitution in Antiquity*.

70. Edge, "Python Gets a New Governance Model."

71. Clark, "DRAG THEM."

72. brown, *We Will Not Cancel Us*, 65 and 54. Perhaps also relevant is the "online disinhibition effect"; see Suler, "The Online Disinhibition Effect."

73. Phillips, *This Is Why We Can't Have Nice Things*, 12.

2. HOMESTEADING ON A SUPERHIGHWAY: HOW THE POLITICS OF NO-POLITICS AIDED AN AUTHORITARIAN REVIVAL

1. Barbrook and Cameron, "The Californian Ideology," 52.

2. Turner, *From Counterculture to Cyberculture*, 4.

3. With *encoded*, I refer to the political logic of Hall, "Encoding/Decoding."

4. Tria Kerkvliet, "Everyday Politics in Peasant Societies (and Ours)."

5. Turner, *From Counterculture to Cyberculture*; Curtis, *All Watched Over by Machines of Loving Grace*; Rankin, *A People's History of Computing in the United States*.

6. Hepp, *Deep Mediatization*; Lofton, *Consuming Religion*, 2.

7. Meehan and Turner, *Seeing Silicon Valley*.

8. Driscoll, *The Modem World*, ch. 5.

9. Rheingold, *The Virtual Community*, 2000, 325.

10. Limerick, *The Legacy of Conquest*; Russo, *The Infinite Machine*.

11. Limerick, *The Legacy of Conquest*, 55.

12. Merk and Merk, *Manifest Destiny and Mission in American History*; Miller, "American Indians, the Doctrine of Discovery, and Manifest Destiny."

13. Barbrook and Cameron, "The Californian Ideology," 58, 61.

14. Rheingold, *The Virtual Community*, 2000, 31

15. Rheingold, "Online Community Governance."

16. Hafner, "The Epic Saga of The Well"; Turner, *From Counterculture to Cyberculture*.

17. Hafner, "The Epic Saga of The Well."

18. Rheingold, "Online Community Governance"; The WELL, "How The Well Bought Itself."

19. hooks, "Homeplace (A Site of Resistance)," 384. Cf. the concept of counterpublics in Black feminist scholarship in Jackson and Foucault Welles, "Hijacking #MYNYPD."

20. Gray, "Negotiating Identities/Queering Desires"; Turner, "The Arts at Facebook."

21. Nozick, *Anarchy, State, and Utopia*.

22. Frey and Schneider, "Effective Voice" explores aspects of exit in online social media, and Mannan and Schneider, "Exit to Community," focuses on exit as a business model.

23. Cross, "Twitter's 'Vox Populi' Is a Lie"; Pandey and Lawler, "Elon Musk on What the First Mars Colony Will Look Like"; Pogue, "Inside the New Right's Next Frontier"; Rubenstein and Supp-Montgomerie, "Somewhere Out There"; Srinivasan, *The Network State*; Turner, "Burning Man at Google"; Wójtowicz and Szocik, "Democracy or What?"

24. On scale and startups, see Graham, "Startup = Growth"; Thiel and Masters, *Zero to One*. Tsing, "On Nonscalability," critiques the aspiration of "scalability" through an anthropological lens; and Gillespie, "Content Moderation, AI, and the Question of Scale," raises complementary challenges for platform design and moderation. For more on scalability, see chapter 3.

25. For general histories of the Silicon Valley economy, see Nicholas, *VC: An American History*; and O'Mara, *The Code*. For a focus on the connection between geographic colonialism and digital capitalism, see Harris, *Palo Alto*.

26. Barbrook and Cameron, "The Californian Ideology," 62.

27. Nuth, "Two Medieval Soteriologies."

28. Barbrook and Cameron, "The Californian Ideology," 50.

29. Alaimo, "How the Facebook Arabic Page 'We Are All Khaled Said' Helped Promote the Egyptian Revolution." Facebook removed the page because he was not using a "real name," until well-connected Silicon Valley operators intervened.

30. Schneider, *Thank You, Anarchy*, 90, 93.

31. Kraidy, "Fun against Fear in the Caliphate," 46.

32. Sharlet, "'He's the Chosen One to Run America.'"

33. Hoback, *Q: Into the Storm*; Kirkpatrick, "Who Is Behind QAnon?"

34. Gillespie, "The Politics of 'Platforms.'"

35. Phillips and Milner, *The Ambivalent Internet*.

36. Johnson and Stokols, "What Steve Bannon Wants You to Read"; Pogue, "Inside the New Right's Next Frontier"; Siegel, "The Red-Pill Prince"; Tait, "Mencius Moldbug and Neoreaction."

37. Rheingold, *The Virtual Community*, 2000, 298

38. brown, *Emergent Strategy*, 328.

39. Certeau, *The Practice of Everyday Life*.

40. brown, *Emergent Strategy*, 88.

41. Sheldrake, "Michel de Certeau"; Lefebvre, *Critique of Everyday Life*.

42. Highmore, *Everyday Life and Cultural Theory*, 150. See also Felski, "The Invention of Everyday Life." For the everyday in media theory, see Bucher, "The Algorithmic Imaginary"; Kember and Zylinska, *Life after New Media*; Papacharissi, *Affective Publics*; Phillips and Milner, *The Ambivalent Internet*.

43. Albergotti, "He Predicted the Dark Side of the Internet 30 Years Ago."

44. Agre, "The Dynamic Structure of Everyday Life"; Agre, "The Structures of Everyday Life."

45. Agre, "The Practical Republic." For a more recent aligned argument, see Zuckerman, "How Social Media Could Teach Us to Be Better Citizens."

46. Agre, *Computation and Human Experience*.

47. Agre, "The Practical Republic."

48. Ramirez, Saucerman, and Dietmeier, "Twitch Plays Pokemon"; Woessner, "Teaching with SimCity."

49. Schneider, *Everything for Everyone*, 93; Schneider, *Thank You, Anarchy*, 64.

50. brown, *Emergent Strategy*, 146.

51. Rheingold, *The Virtual Community*, 2000, 395.

52. Personal communication, February 16 and 18, 2019. I am not sure whether the correspondent was who they said they were.

53. Sherman, "Don't Be Fooled by Big Tech's Anti-China Sideshow."

54. Stasavage, *The Decline and Rise of Democracy*.

55. Tomba, "Of Quality, Harmony, and Community"; Wang, Juffermans, and Du, "Harmony as Language Policy in China."

56. Tang, *Chinese Modern*, 290.

57. Luo and Lu, "Participatory Censorship"; Ng, *Blocked on Weibo*.

58. Tria Kerkvliet, "Everyday Politics in Peasant Societies (and Ours)," 232.

59. Havel, "The Power of the Powerless."

60. Barbrook and Cameron, "The Californian Ideology," 63.

3. DEMOCRATIC MEDIUMS: CASE STUDIES IN POLITICAL IMAGINATION

1. For example, Applebaum and Pomerantsev, "How to Put Out Democracy's Dumpster Fire"; Vaidhyanathan, *Antisocial Media*.

2. Specifically, she used it as the title of the first chapter of Guinier, *Lift Every Voice*. The phrase was coined in an article about her in *Political Woman* magazine.

3. Guinier, *Lift Every Voice*, 126; Will, "Sympathy For Guinier."

4. Guinier, *Lift Every Voice*, 37.

5. Guinier and Torres, *The Miner's Canary*.

6. Stout, *Democracy and Tradition*.

7. Allen, *Tocqueville, Covenant, and the Democratic Revolution*, 12; Tocqueville, *Democracy in America*.

8. Allen, *Tocqueville, Covenant, and the Democratic Revolution*, xiv. On civil religion more generally, see Bellah, "Civil Religion in America." Norton, *Wild Democracy*, 64, summarizes the Tocquevillian view this way: "People who rule themselves reach upward to the divine."

9. Goodman, "A Year after 'Defund,' Police Departments Get Their Money Back"; Gramlich, "Violent Crime Is a Key Midterm Voting Issue, but What Does the Data Say?"

10. Kaba, "Yes, We Mean Literally Abolish the Police." On the movement more generally, see Cullors, *An Abolitionist's Handbook*; Dixon and Piepzna-Samarasinha, *Beyond Survival*; Kaba and Ritchie, *No More Police*; Smith, "Abolition Feminism and Jumping Scale."

11. Creative Interventions, *Creative Interventions Toolkit*.

12. Interrupting Criminalization and Project Nia, One Million Experiments.

13. Kaba and Hassan, *Fumbling towards Repair*.

14. Dixon and Piepzna-Samarasinha, *Beyond Survival*, "Building Community Safety."

15. Creative Interventions, *Creative Interventions Toolkit*, sec. 3, p. 32.

16. Johnson, "Why Violent Crime Surged after Police across America Retreated." In this case it is also worth considering resurgent racism, the force that led to the demise of Reconstruction many decades earlier, according to Du Bois, *Black Reconstruction in America*.

17. Vaughn, Peyton, and Huber, "Mass Support for Proposals to Reshape Policing Depends on the Implications for Crime and Safety."

18. Kaba and Hassan, *Fumbling towards Repair*, "Opening Thoughts." Other examples of the conversational quality of the literature include Dixon and Piepzna-Samarasinha, *Beyond Survival*; Kaba and Murakawa, *We Do This 'til We Free Us*; brown, *Emergent Strategy*; brown, *Pleasure Activism*.

19. brown, *Emergent Strategy*; brown and Imarisha, *Octavia's Brood*.

20. Davis, *Abolition Democracy*, 91–92; Du Bois, *Black Reconstruction in America*. For more on Du Bois's understanding of abolition and economic democracy, see Du Bois, "Of the Ruling of Men."

21. Quoted in Kaba and Ritchie, *No More Police*, 16, 247.

22. brown, "What Is/Isn't Transformative Justice?" 389.

23. Kaba and Hassan, *Fumbling towards Repair*, "Who Is This For?"

24. Kaba, "Be Humble," 455.

25. On intersectional experiences of online harm, see Musgrave, Cummings, and Schoenebeck, "Experiences of Harm, Healing, and Joy among Black Women and Femmes on Social Media." On LambdaMOO, see Dibbell, "A Rape in Cyberspace."

26. From Boggs and Kurashige, *The Next American Revolution*, quoted in Kaba and Ritchie, *No More Police*, 211.

27. Boggs and Kurashige, *The Next American Revolution*, 50. Boggs credits systems thinker Margaret J. Wheatley for this phrasing.

28. Hasinoff and Schneider, "From Scalability to Subsidiarity in Addressing Online Harm." See also Hasinoff, Gibson, and Salehi, "The Promise of Restorative Justice in Addressing Online Harm."

29. Tsing, "On Nonscalability." See also her book-length treatment, Tsing, *The Mushroom at the End of the World*.

30. Haldane, "On Being the Right Size."

31. Lippmann, "The Basic Problem of Democracy"; Lippmann, *Public Opinion*.

32. Graham, "Startup = Growth"; Sullivan, "Blitzscaling."

33. Gillespie, "Content Moderation, AI, and the Question of Scale," 3.

34. Follesdal, "Survey Article: Subsidiarity"; Brennan, "Subsidiarity in the Tradition of Catholic Social Doctrine." For an application to blockchain systems, see Siddarth, Allen, and Weyl, "The Web3 Decentralization Debate Is Focused on the Wrong Question."

35. For studies on purchasing cooperatives in particular, see Martins Rodrigues and Schneider, "Scaling Co-operatives through a Multi-Stakeholder Network"; Taylor, "An

Analysis of the Entrepreneurial Institutional Ecosystems Supporting the Development of Hybrid Organizations."

36. Katsaros et al., "Procedural Justice and Self-Governance on Twitter," 12. For additional evidence that more attenuated, contextual procedures improve perceptions of legitimacy, see Fan and Zhang, "Digital Juries"; McGillicuddy, Bernard, and Cranefield, "Controlling Bad Behavior in Online Communities"; Pan et al., "Comparing the Perceived Legitimacy of Content Moderation Processes." On conceptions of scale in transformative justice, see Smith, "Abolition Feminism and Jumping Scale."

37. This typology derives from Jhaver, Frey, and Zhang, "Designing for Multiple Centers of Power." We apply it to subsidiarity specifically in Hasinoff and Schneider, "From Scalability to Subsidiarity in Addressing Online Harm." On federalism generally, see Bednar, *The Robust Federation*. On polycentricity generally, see Ostrom, "Beyond Markets and States."

38. Caelin, "Decentralized Networks vs. the Trolls." Mastodon is considered an example of a "federated social network" and part of the "fediverse," but in terms of the typology presented here, it is more accurately described as polycentric rather than federated. For a less hopeful account, see Bevensee and Rebellious Data LLC, "The Decentralized Web of Hate."

39. See Mia Mingus, "Pods and Pod-Mapping Worksheet," in Dixon and Piepzna-Samarasinha, *Beyond Survival*, a framework that avoids the language of "community" with more specifically defined "pods" in an accountability process.

40. Dickie, *Hollow Water*.

41. Alexander, "Meditations on Moloch."

42. Gray, "Transgression, Release and 'Moloch'"; Marx, *Capital*, 397; Marx, *Grundrisse*, 199.

43. MolochDAO, The Original Grant Giving DAO. See also Soleimani et al., "The Moloch DAO."

44. MolochDAO, The Original Grant Giving DAO.

45. Hito Steyerl, "Walk the Walk—Beyond Blockchain Orientalism," in Catlow and Rafferty, *Radical Friends*.

46. bendi, "Introducing Flexible Voting"; Buterin, "Moving beyond Coin Voting Governance"; Emmett, "Conviction Voting"; Wright, "Quadratic Voting and Blockchain Governance"; Reijers et al., "Now the Code Runs Itself"; Zargham, "Sensor Networks and Social Choice."

47. For a review of blockchain-based identity projects, see Siddarth et al., "Who Watches the Watchmen?"

48. Babbitt and Dietz, "Crypto-Economic Design"; Brekke and Alsindi, "Cryptoeconomics"; Buterin, *Proof of Stake*; Stark, "Making Sense of Cryptoeconomics."

49. Buterin, "Governance, Part 2."

50. For an anthropological perspective on this process, see Maurer, Nelms, and Swartz, "'When Perhaps the Real Problem Is Money Itself!'"

51. Korpas, Frey, and Tan, "Political, Economic, and Governance Attitudes of Blockchain Users." The survey was developed through the Metagovernance Project, a research collaboration that I co-lead, though I was not directly involved in this specific effort.

52. Golumbia, *The Politics of Bitcoin*; Brunton, *Digital Cash*.

53. Swartz, "Blockchain Dreams"; Schneider, "Decentralization: An Incomplete Ambition."

54. A further dimension of distributed design appears in application-specific blockchains, as in the Cosmos network, or in user-specific chains, as in Holochain.

55. On dynamic decision-making, see Beck, Müller-Bloch, and King, "Governance in the Blockchain Economy"; Reijers et al., "Now the Code Runs Itself." On voting systems, see Emmett, "Conviction Voting"; Karjalainen, "Governance in Decentralized Networks"; Zargham, "Sensor Networks and Social Choice." On incentive alignment, see Beck, Müller-Bloch, and King, "Governance in the Blockchain Economy"; Karjalainen, "Governance in Decentralized Networks." On dispute resolution, see Barnett and Treleaven, "Algorithmic Dispute Resolution of Professional Dispute Resolution Using AI and Blockchain Technologies"; Aouidef, Ast, and Deffains, "Decentralized Justice." On permissionlessness, see Beck, Müller-Bloch, and King, "Governance in the Blockchain Economy." On shared ownership, see Beck, Müller-Bloch, and King, "Governance in the Blockchain Economy"; Fritsch, "The Common Factory"; Mannan and Schneider, "Exit to Community." On security and censorship, see De Filippi, Mannan, and Reijers, "Blockchain as a Confidence Machine." On sovereignty, see Duffy, "The Time for Self-Sovereign Identity Is Now"; Manski and Manski, "No Gods, No Masters, No Coders?"; De Filippi and Wright, *Blockchain and the Law*. On transparency, see Cila et al., "The Blockchain and the Commons." On governance markets, see Alston, "Constitutions and Blockchains." On exit, see Caliskan, "Data Money Makers"; Ba, Zignani, and Gaito, "The Role of Groups in a User Migration across Blockchain-Based Online Social Media." On identity, see Allen, "The Path to Self-Sovereign Identity"; Siddarth et al., "Who Watches the Watchmen?" On interfaces, see, e.g., Snapshot, snapshot.org; Tally, withtally.com.

56. Schneider, "Mediated Ownership"; Swartz, *New Money*.

57. Cuende quoted in Aragon, "The Fight for Freedom."

58. This argument is elaborated in Schneider, "Cryptoeconomics as a Limitation on Governance." See also Vitalik Buterin's response, Buterin, "On Nathan Schneider on the Limits of Cryptoeconomics."

59. Brown, *Undoing the Demos*, 22. See also Hayden, *The Zapatista Reader*; and Davies, *The Limits of Neoliberalism*.

60. Arendt, *The Human Condition*. See also Arndt, *Arendt on the Political*; Klein, "'Fit to Enter the World.'"

61. Satz, *Why Some Things Should Not Be for Sale*; Bowles, *The Moral Economy*. On Homo speculans, see Komporozos-Athanasiou, *Speculative Communities*.

62. Alston, "Political and Economic Institutional Emergence."

63. For instance, Buterin, "Governance, Part 2"; Buterin, "Blockchain Voting Is Overrated among Uninformed People but Underrated among Informed People"; Ferreira, Li, and Nikolowa, "Corporate Capture of Blockchain Governance."

64. Ferraro, Pfeffer, and Sutton, "Economics Language and Assumptions," 14.

65. Gritsenko and Wood, "Algorithmic Governance," 1.

66. MacKinnon, "What to Get Right First," eloquently warns of the danger crypto presents and the need to anticipate abuses in system designs. I proposed possible strategies in Schneider, "How We Can Encode Human Rights in the Blockchain."

67. MolochDAO, The Original Grant Giving DAO; Buterin, Hitzig, and Weyl, "Liberal Radicalism"; Schneider, "An Economy for Anything."

68. Conte de Leon et al., "Blockchain: Properties and Misconceptions"; Siddarth et al., "Who Watches the Watchmen?"

69. For instance, Aouidef, Ast, and Deffains, "Decentralized Justice"; Eth_man, "Everything about 1Hive in One Place"; Wright, "Quadratic Voting and Blockchain Governance";

El Faqir, Arroyo, and Hassan, "An Overview of Decentralized Autonomous Organizations on the Blockchain."

70. Buterin, "Soulbound"; Weyl, Ohlhaver, and Buterin, "Decentralized Society." For the longer trajectory of his thinking, see Buterin, *Proof of Stake*.

71. For instance, Atzori, "Blockchain Technology and Decentralized Governance"; Clifton and Pal, "The Policy Dilemmas of Blockchain"; COALA, "Model Law for Decentralized Autonomous Organizations"; Hinman, "Digital Asset Transactions." For counterpoints, see Alston, "Blockchain and the Law"; De Filippi, Mannan, and Reijers, "The Alegality of Blockchain Technology."

72. Schneider, *Everything for Everyone*; Kreutler, "A Prehistory of DAOs"; Walden, "Past, Present, Future."

73. Ahonen, "Rescuing Crypto Workers from Terrible US Job Conditions"; Radebaugh and Muchnik, "Solving the Riddle of the DAO with Colorado's Cooperative Laws." See also FairCoin and experimental cryptocurrency (König et al., "The Proof-of-Cooperation Blockchain FairCoin") and DisCO, a feminist-informed framework for cooperative DAOs (Troncoso and Utratel, *Groove Is in the Heart*).

74. Alston, "Constitutions and Blockchains"; Tan et al., "Constitutions of Web3." For an extended discussion of human rights and blockchains, see Schneider, "How We Can Encode Human Rights in the Blockchain." I am grateful for Ethan Zuckerman for helping me recognize the centrality of rights to democratic governance.

75. For a review of cryptoeconomic enforcement and dispute-resolution mechanisms, see Aouidef, Ast, and Deffains, "Decentralized Justice." For non-crypto examples of online juries, see Kou et al., "Managing Disruptive Behavior through Non-Hierarchical Governance"; Johansson, Verhagen, and Kou, "I Am Being Watched by the Tribunal."

76. Swartz, "Theorizing the 2017 Blockchain ICO Bubble as a Network Scam." See also web3isgoinggreat.com.

77. Buterin, "Blockchain Voting Is Overrated among Uninformed People but Underrated among Informed People."

78. For further evidence of the devotion to chess, see "Christmas Special," in Buterin, *Proof of Stake*.

79. Dailey, "Vitalik Buterin Says."

80. Huizinga, *Homo Ludens*. This is thanks to crypto entrepreneur Joel Dietz, who suggested it in advance of a meeting of the Metagovernance Project's weekly seminar, which I then co-organized. For examples of gamification in recent experiments in democratic innovation, see Hon, "How Game Design Principles Can Enhance Democracy."

81. Huizinga, *Homo Ludens*, 85.

82. brown, *Pleasure Activism*, 23.

83. Ostrom, "Artisanship and Artifact," 310; Sabetti, "Constitutional Artisanship and Institutional Diversity"; Fotos, "Vincent Ostrom's Revolutionary Science of Association." I am grateful to Keith Taylor for guiding me to this side of Ostrom's thought.

4. GOVERNABLE STACKS: ORGANIZING AGAINST DIGITAL COLONIALISM

1. James, *The Black Jacobins*, 267.

2. Bird et al., *Finally Got the News*.

3. On digital colonialism, see Jandrić and Kuzmanić, "Digital Postcolonialism"; Pinto, "Digital Sovereignty or Digital Colonialism?"; Kwet, "Digital Colonialism"; Avila, "Against Digital Colonialism." On technocolonialism, see Madianou, "Technocolonialism." On data colonialism, see Thatcher, O'Sullivan, and Mahmoudi, "Data Colonialism through Accumulation by Dispossession"; Couldry and Mejias, "Making Data Colonialism Liveable." On data orientalism, see Kotliar, "Data Orientalism." On digital capitalism, see Qiu, *Goodbye iSlave*. On digital extractivism, see Iyer et al., "Automated Imperialism, Expansionist Dreams." On platform imperialism, see Jin, "The Construction of Platform Imperialism in the Globalization Era." On postcolonial computing, see Irani et al., "Postcolonial Computing." On decolonial computing, see Ali, "A Brief Introduction to Decolonial Computing." On imperial play, see Merwe, "Imperial Play."

4. Ali, "A Brief Introduction to Decolonial Computing"; Harney and Moten, *The Undercommons*, 87.

5. Cooper, *Colonialism in Question*; Tuck and Yang, "Decolonization Is Not a Metaphor."

6. Thaning, Gudmand-Høyer, and Raffnsøe, "Ungovernable: Reassessing Foucault's Ethics"; Anderson, "Ungovernable: An Interview with Lorenzo Kom'boa Ervin"; Li, "Governmentality."

7. Castellanos and Pine, "Berta Cáceres in Her Own Words."

8. This observation builds on Paul, "Antitrust as Allocator of Coordination Rights," discussed further in chapter 5.

9. Benjamin, *Race after Technology*; Eubanks, *Automating Inequality*; Madianou, "Technocolonialism"; Couldry and Mejias, "Data Colonialism"; Kwet, "Digital Colonialism"; Milan and Treré, "Big Data from the South(s)"; Duarte et al., "'Of Course, Data Can Never Fully Represent Reality'"; Tufekci, *Twitter and Tear Gas*; Canella, "Racialized Surveillance."

10. Garza, *The Purpose of Power*, 273.

11. Harney and Moten, *The Undercommons*, 56, 55, 57, 20. For an example of "governance" used in apparent alignment with their meaning, see the recent RAND Corporation report Frank et al., "Adaptive Engagement for Undergoverned Spaces."

12. Fanon, *The Wretched of the Earth*, 81.

13. Lebrón Ortiz, "Resisting (Meta) Physical Catastrophes through Acts of Marronage."

14. Harney and Moten, *The Undercommons*, 155.

15. James, Lee, and Chaulieu, *Facing Reality*, 121–22. This method resembles the "rough consensus" of Internet protocol governance; see, for instance, Resnick, "On Consensus and Humming in the IETF."

16. Hardt and Negri, *Assembly*; Castells, *Networks of Outrage and Hope*.

17. brown, *Emergent Strategy*; Invisible Committee, *The Coming Insurrection*; Vitalist International, "Life Finds a Way."

18. Lenin, "What Is To Be Done?"

19. Luxemburg, "Organizational Questions of Russian Social Democracy."

20. Fanon, *The Wretched of the Earth*, 130, 142.

21. Fitzpatrick, "'You Never Know When It's Going to Explode'"; Johnson, "They Showed the Way to Labor Emancipation!"; James, "Every Cook Can Govern." "Johnson" is a pseudonym for James.

22. Getachew, *Worldmaking after Empire*.

23. Boggs, *Living for Change*; Boggs and Kurashige, *The Next American Revolution*; Boggs and Boggs, *Revolution and Evolution in the Twentieth Century*; King, "Introduction to Boggs."

24. Nunes, *Neither Vertical nor Horizontal*, ch. 1, conclusion.

25. Boggs and Kurashige, *The Next American Revolution*, 68

26. Harney and Moten, *The Undercommons*, 153.

27. Donovan, Dreyfuss, and Friedberg, *Meme Wars*; Megarry, *The Limitations of Social Media Feminism*; Papacharissi, *Affective Publics*; Tufekci, *Twitter and Tear Gas*.

28. Schneider, *Thank You, Anarchy*, 36.

29. Schneider, "How OWS' 'Anti-Market Research Analyst' Helps the Movement Go Viral."

30. Tufekci, *Twitter and Tear Gas*.

31. Dean, "Communicative Capitalism"; Kazan, *On the Waterfront*.

32. Desai, "Charlie Chaplin Meets Gandhiji"; Chaplin, *My Autobiography*, 373.

33. Chaplin, *My Autobiography*, 415. On the connection between looms and early computing through the work of Ada Lovelace and Charles Babbage, see Park and Jayaraman, "Textiles and Computing."

34. Rao, "Gandhi, Untouchability and the Postcolonial Predicament"; Roy, *The Doctor and the Saint*; Mishra, "Gandhi for the Post-Truth Age."

35. Gandhi, *Constructive Programme*, 29.

36. Brown, "Spinning without Touching the Wheel"; Fuchs, "M. N. Roy and the Frankfurt School," 257; Garg and Camp, "Gandhigiri in Cyberspace."

37. Vincent, "Boards That Gave Artisans a Voice Scrapped."

38. Haraway, "A Cyborg Manifesto"; Puar, "'I Would Rather Be a Cyborg Than a Goddess.'" See also my lab's pamphlet on "cyborg community," Argast et al., *Sacred Stacks*.

39. Bratton, *The Stack*, xvii.

40. On May First Movement Technology, see Lopez et al., *The Organic Internet*; Schneider, *Everything for Everyone*, ch. 5; Schneider, "The Joy of Slow Computing." On Indymedia, see Pickard, "United yet Autonomous"; Wolfson, *Digital Rebellion*.

41. Bowles, "A Dark Consensus about Screens and Kids Begins to Emerge in Silicon Valley"; Marantz, "Silicon Valley's Crisis of Conscience."

42. Ehmke, "A Six-Month Retrospective on Ethical Open Source"; Schneider, "The Tyranny of Openness"; SSL Nagbot, "Feminist Hacking/Making."

43. Tarnoff, *The Making of the Tech Worker Movement*. See also the Tech Workers Coalition newsletter at news.techworkercoalition.org.

44. Tuck and Yang, "Decolonization Is Not a Metaphor," 36.

45. Murray, "A 'Feminist' Server to Help People Own Their Own Data."

46. Zuckerman, "Cute Cats to the Rescue?," suggests that platforms designed for mass appeal are vital for online political strategy. These could be governable stacks but currently they are not.

47. brown, *Emergent Strategy*, 90. In the original, this passage is italicized.

48. Cong, "Contesting Freedom of Information"; Freuler, "The Case for a Digital Non-Aligned Movement"; Mejias, "To Fight Data Colonialism, We Need a Non-Aligned Tech Movement."

49. Bevensee and Rebellious Data LLC, "The Decentralized Web of Hate"; Parler, "What Is the Parler Community Jury?"

50. These are explored in more depth, with numerous examples, in Schneider, "Governable Stacks Against Digital Colonialism." For another example of an assessment framework for ethical technology choices, see Bogaerts, Van Dijck, and Zuckerman, "Creating Public-Spaces."

51. Terranova, "Red Stack Attack!" On the maintenance and use of legacy technologies, see Edgerton, *The Shock of the Old*; Maxigas and Latzko-Toth, "Trusted Commons"; Roscam Abbing, "On Cultivating the Installable Base."

52. Kelty, *Two Bits*. See also a related concept to "recursive publics" in the more recent Nabben, "Web3 as 'Self-Infrastructuring.'"

53. Nucera and Mogilevich, "Teaching Community Technology Handbook," 104.

54. Here I refer to Sans Souci Cooperative, one of the participating organizations in my lab's Sacred Stacks project: colorado.edu/lab/medlab/sacred-stacks.

55. Schneider et al., "Modular Politics." Notably, the Metagovernance Project first emerged from a consulting project (led by legal scholar Lawrence Lessig) with a game company looking to incorporate self-governance among players.

56. Quoted from Schneider et al., "Modular Politics."

57. Schneider and Hasinoff, "Mastodon Isn't Just a Replacement for Twitter."

58. In political theory, see Alston, "Constitutions and Blockchains"; in cybernetic theory, see Zargham and Nabben, "Aligning 'Decentralized Autonomous Organization' to Precedents in Cybernetics."

59. Hwang and Shaw, "Rules and Rule-Making in the Five Largest Wikipedias."

60. Ostrom, *Understanding Institutional Diversity*, 7. See also the contention in Murimi, "Governance in DAOs," that modular, composable governance occurs in parallel ways among primates and DAOs.

61. Zhang, Hugh, and Bernstein, "PolicyKit"; see also gateway.metagov.org.

62. Barandiaran et al., "Decidim."

63. Kreutler, "Zodiac"; Aragon, "Introducing aragonOS"; Webber, "Introducing OpenZeppelin Governor."

64. Ingersoll, "Wampum and Its History"; Yang, *Cowrie Shells and Cowrie Money*. Hartman, *Lose Your Mother*, 206–10, in particular alerted me to this legacy. Winger-Bearskin, "Before Everyone Was Talking about Decentralization, Decentralization Was Talking to Everyone," makes the connection between wampum belts (which in some contexts were made with cowrie shells) and blockchains. Her contributions to the Excavations artist cohort (profiled in this book) were vital to my thinking about ancestry in the context of governance archaeology. COWRIE is also the name of a South African cryptocurrency, and there is a crypto consultancy called Cowrie LLC. Cordes, "Storying Indigenous Cryptocurrency," explores intersections of shell economies and crypto.

65. Norton, *Wild Democracy*, 63. Norton's approach also challenges the premises of governance archaeology when she writes later on, "We do not need any past at all.... Whatever the place, whenever the time, whoever the people: democracy can begin there" (153).

66. Jacobs, "Iroquois Great Law of Peace and the United States Constitution"; Payne, "The Iroquois League, the Articles of Confederation, and the Constitution."

67. Carugati and Schneider, "Governance Archaeology: Research as Ancestry." I understand governance archaeology as a kind of accompaniment to media archaeology. See, for instance, Huhtamo and Parikka, *Media Archaeology*; and Sengupta, "Towards a Decolonial Media Archaeology." It also reflects the postcolonial turn in "regular" archaeology, such as Londoño, "Indigenous Archaeology, Community Archaeology, and Decolonial Archaeology"; and research methodologies in general, such as Smith, *Decolonizing Methodologies: Research and Indigenous Peoples*.

68. Harari, *Sapiens*; Graeber and Wengrow, *The Dawn of Everything*. The debate is in some sense a recapitulation of the retort to social Darwinism in Kropotkin, *Mutual Aid*. Carugati and I are especially influenced by another "big history," Stasavage, *The Decline and Rise of Democracy*.

69. Quoted in Vázquez, "Towards a Decolonial Critique of Modernity."

70. Sousa Santos, *Epistemologies of the South*. See also Quijano, "Coloniality and Modernity/Rationality," whose vision of a post-colonial "totality" motivates our efforts of juxtaposition in a database.

71. Stasavage, *The Decline and Rise of Democracy*, is particularly generous with examples of this sort of thing.

72. Vázquez, "Towards a Decolonial Critique of Modernity."

73. See, for example, Audrey Huntley's chapter, "From Breaking Silence to Community Control," in Dixon and Piepzna-Samarasinha, *Beyond Survival*.

5. GOVERNABLE SPACES: DEMOCRACY AS A POLICY STRATEGY

1. Turnbull, "Monsters Come Howling in Their Season."

2. brown and Imarisha, *Octavia's Brood*, 509. See also Butler, *Critical Black Futures*.

3. brown, *Pleasure Activism*, 17.

4. See, for instance, Benjamin, *Race after Technology*; Eubanks, *Automating Inequality*.

5. On my work in the cooperative movement, for instance, I co-organized the first Platform Cooperativism conference at the New School in 2015, co-founded the accelerator Start.coop, and serve on the board of Zebras Unite, a feminist network organizing a startup ecosystem through cooperatives. See Schneider, *Everything for Everyone*; and Scholz and Schneider, *Ours to Hack and to Own*, for more context. On the partner state, see Bauwens, Kostakis, and Pazaitis, *Peer to Peer*; Restakis, *Civilizing the State*.

6. Gillespie, "Governance of and by Platforms"; Klonick, "The New Governors"; Lessig, *Code*.

7. Berners-Lee, *Weaving the Web*; Gillespie, "The Politics of 'Platforms'"; Milan and Treré, "Big Data from the South(s)"; Nash, "International Facebook 'Friends'"; Schneider, "Democrats and Republicans Both Have a Big Blind Spot on Net Neutrality."

8. Freeman, "The Tyranny of Structurelessness"; SSL Nagbot, "Feminist Hacking/Making."

9. Arcy, "Emotion Work"; Lewis, Rowe, and Wiper, "Online Abuse of Feminists as an Emerging Form of Violence against Women and Girls"; Musgrave, Cummings, and Schoenebeck, "Experiences of Harm, Healing, and Joy among Black Women and Femmes on Social Media"; Schoenebeck, Haimson, and Nakamura, "Drawing from Justice Theories to Support Targets of Online Harassment"; Schor, *After the Gig*; SSL Nagbot, "Feminist Hacking/Making."

10. Buolamwini and Gebru, "Gender Shades"; Browne, *Dark Matters*; Noble, *Algorithms of Oppression*.

11. Klonick, "The New Governors"; Plantin and Seta, "WeChat as Infrastructure"; Rochefort, "Regulating Social Media Platforms."

12. Cady, "Flexible Labor"; Federici, *Revolution at Point Zero*; Jarrett, "The Relevance of 'Women's Work'"; Nakamura, "The Unwanted Labour of Social Media"; Rankin, *A People's*

History of Computing in the United States; Rochefort, "Regulating Social Media Platforms"; Schneider, "The Tyranny of Openness."

13. Cohen, "A 1970s Essay Predicted Silicon Valley's High-Minded Tyranny"; Freeman, "The Tyranny of Structurelessness."

14. Finley, "The Woman Bringing Civility to Open Source Projects"; Savic and Wuschitz, "Feminist Hackerspace as a Place of Infrastructure Production"; Toupin, "Feminist Hackerspaces."

15. Kember and Zylinska, *Life after New Media*; Zobl and Drüeke, *Feminist Media*.

16. Association for Progressive Communications, "Feminist Principles of the Internet."

17. Gurevich, "Patriarchy?," 518.

18. Kosseff, *The Twenty-Six Words That Created the Internet*.

19. Kosseff, *The Twenty-Six Words That Created the Internet*, 248.

20. Klonick, "The New Governors"; Lemley, "The Splinternet."

21. Citron and Wittes, "The Internet Will Not Break."

22. Casarosa, "Transnational Collective Actions for Cross-Border Data Protection Violations"; Seering, "Reconsidering Self-Moderation."

23. Aristotle, *Nicomachean Ethics*, bk. VIII; Nordin, "Decolonising Friendship."

24. Friedman, "Feminism and Modern Friendship," 286. See also Devere, "The Academic Debate on Friendship and Politics"; Devere and Smith, "Friendship and Politics"; Dillard, "To Experience Joy"; Schwarzenbach, "Democracy and Friendship."

25. Fiesler, Morrison, and Bruckman, "An Archive of Their Own"; Megarry, *The Limitations of Social Media Feminism*; Toupin, "Feminist Hackerspaces."

26. djahnie, "How Does the Conflict Resolution Feature Work?"; Friedman, "Autonomy and Social Relationships."

27. Paul, "Antitrust as Allocator of Coordination Rights."

28. Federici, *Caliban and the Witch*; Federici, *Revolution at Point Zero*.

29. Johnston and Land-Kazlauskas, "Organizing On-Demand"; Vaheesan and Schneider, "Cooperative Enterprise as an Antimonopoly Strategy."

30. Schor, *After the Gig*.

31. Federici, *Revolution at Point Zero*; Jarrett, "The Relevance of 'Women's Work'"; Waring, *If Women Counted*.

32. SSL Nagbot, "Feminist Hacking/Making"; Rankin, *A People's History of Computing in the United States*.

33. Arcy, "Emotion Work"; Illich, *Shadow Work*; Jarrett, "The Relevance of 'Women's Work'"; SSL Nagbot, "Feminist Hacking/Making"; Terranova, "Free Labor."

34. Gray and Suri, *Ghost Work*; Irani, "Difference and Dependence among Digital Workers."

35. Schor, *After the Gig*, 2.

36. Gonzalez, AB-1319.

37. Fisk, "Hollywood Writers and the Gig Economy."

38. Lozano-Paredes, "Emergent Transportation 'Platforms' in Latin America."

39. Schneider, "The Future of Owning the Internet."

40. Galperin and Bar, "The Microtelco Opportunity."

41. Thompson, Gómez, and Toro, "Women's Alternative Internet Radio and Feminist Interactive Communications"; Serafini, "Community Radio as a Space of Care"; Arriola, "The Feminist Radio Collective of Peru."

42. Harding, *The Science Question in Feminism*; Russ, "SF and Technology as Mystification"; Supp-Montgomerie, "Infrastructural Awareness"; Wajcman, *Feminism Confronts Technology*.

43. Parks, "Around the Antenna Tree"; Burrington, *Networks of New York*.

44. Edwards et al., "Understanding Infrastructure"; Parks, "'Stuff You Can Kick.'"

45. Ali, *Farm Fresh Broadband*; Doyle, *Lines across the Land*; Oakland, "Minnesota's Digital Divide"; Schneider, *Everything for Everyone*, ch. 6; Talbot, Hessekiel, and Kehl, "Community-Owned Fiber Networks."

46. Galloway, *Protocol*; Plantin et al., "Infrastructure Studies Meet Platform Studies in the Age of Google and Facebook."

47. Nextcloud, "German Federal Administration Relies on Nextcloud as a Secure File Exchange Solution"; Hodgson, "Matrix and Riot Confirmed as the Basis for France's Secure Instant Messenger App"; Lynch, "Contesting Digital Futures."

48. Detroit Digital Justice Coalition, *How to DiscoTech*; Toupin, "Feminist Hackerspaces."

49. Schoenebeck, Haimson, and Nakamura, "Drawing from Justice Theories to Support Targets of Online Harassment."

50. Feiner and Rodriguez, "Mark Zuckerberg."

51. Strober, "Rethinking Economics through a Feminist Lens."

52. Power, "Social Provisioning as a Starting Point for Feminist Economics."

53. Łapniewska, "Reading Elinor Ostrom through a Gender Perspective"; May and Summerfield, "Creating a Space Where Gender Matters."

54. Hess and Ostrom, "A Framework for Analyzing the Knowledge Commons"; Ostrom, *Governing the Commons*; Ostrom, *Understanding Institutional Diversity*.

55. Chwalisz, "A Movement That's Quietly Reshaping Democracy for the Better"; Giraudet et al., "'Co-construction' in Deliberative Democracy."

56. Chwalisz, *Innovative Citizen Participation and New Democratic Institutions*; Landemore, *Open Democracy*; Procaccia, "Citizens' Assemblies Are Upgrading Democracy." For a critique of assemblies as demobilizing and depoliticizing, see Machin, "Democracy, Agony, and Rupture."

57. Giraudet et al., "'Co-construction' in Deliberative Democracy," 4. On the importance of institutional stakes, see Boswell, Dean, and Smith, "Integrating Citizen Deliberation into Climate Governance."

58. Sousa Santos, "Participatory Budgeting in Porto Alegre"; Bria, "Barcelona Digital City"; Aragón et al., "Deliberative Platform Design"; Barandiaran et al., "Decidim"; Fung and Wright, *Deepening Democracy*.

59. Cowen, "Audrey Tang on the Technology of Democracy"; Hsiao et al., "vTaiwan"; Leonard, "How Taiwan's Unlikely Digital Minister Hacked the Pandemic." The idea of scaling up "our ability to listen" also appeared on public materials for the platform Polis, which Tang has employed in her vTaiwan program. See Computational Democracy Project, "Polis"; Megill, "Pol.is in Taiwan."

60. Bernal, "How Constitutional Crowdsourcing Can Enhance Legitimacy in Constitution Making"; Russell, "Beyond the Local Trap."

61. Srinivasan, *The Network State*. See also Schneider, "Lighten the Load of the Nation-State."

62. Bonneau et al., "Democracy Theatre." On a more recent experiment on Facebook around policies for climate misinformation, see Newton, "Facebook's Big New Experiment in Governance."

63. Matney, "Twitter's Decentralized Future"; Nover, "Jack Dorsey Texted Elon Musk to Say Twitter Never Should Have Been a Company."

64. Burr to Fields, "Re: Concept Release," October 11, 2018; Chesnut to Fields, "Re: Request for Comment," September 21, 2018; Robbins, Schlaefer, and Lutrin, "From Home Sharing and Ride Sharing to Shareholding."

65. Mannan and Schneider, "Exit to Community"; Schneider, "Startups Need a New Option." For more resources on the topic, see colorado.edu/lab/medlab/exit-to-community.

66. Brandel et al., "Sex & Startups." This is an essay by the founders of Zebras Unite, on whose board I serve and which has helped advance the E2C idea.

67. Mannan and Schneider, "Exit to Community"; Pentzien, "Political and Legislative Drivers and Obstacles for Platform Cooperativism in the United States, Germany, and France"; Schneider, "Digital Kelsoism"; Schneider, "Enabling Community-Owned Platforms"; Spicer, "Cooperative Enterprise at Scale."

68. Douek, "Facebook's Oversight Board"; Airbnb, "How We're Giving Hosts a Seat at the Table"; Ovadya, "Towards Platform Democracy."

69. Jäger, Noy, and Schoefer, "What Does Codetermination Do?"

70. Boiron, "Sufficient Decentralization"; Hinman, "Digital Asset Transactions."

71. Nabben et al., "Grounding Decentralised Technologies in Cooperative Principles"; Owocki, "A Brief History of Gitcoin from 2017–2022."

72. Turnbull, "Monsters Come Howling in Their Season."

73. Doyle, *Lines across the Land*; Goodwyn, *The Populist Moment*; Gordon Nembhard, *Collective Courage*; Mayo, *A Short History of Co-operation and Mutuality*; Schneider, *Everything for Everyone*.

EPILOGUE: METAGOVERNANCE

1. Native-Land.ca is produced by Native Land Digital; see Temprano, "Native Land."

2. Coulthard and Hern, "How Flags Divide Us"; Hern is a cooperative developer; Coulthard has pointed to cooperatives as a strategy against "the colonial politics of recognition." See Coulthard, *Red Skin, White Masks*, 68.

3. Butler, "Newhampton"; Gipson, "Afrofuturism's Musical Princess Janelle Monae." For instance, districtox is a network built on Ethereum with Aragon (named after a medieval kingdom in modern Spain) and the Interplanetary File System. Planetary is a decentralized social media network built on Secure Scuttlebutt, which in turn traces its origin story to the open ocean. The examples go on. For a representative example of crypto-post-nationalism, see Aragon, "The Fight for Freedom."

4. brown, "Post Nationalism in the Age of Cooptation and Other Dumpster Fires."

5. Jessop, "Governance and Metagovernance," 5; and Jessop, "The Rise of Governance and the Risks of Failure." See also Braman, "The Irony of Internet Governance Research"; Kooiman, *Modern Governance*; Oever, "The Metagovernance of Internet Governance"; Sørensen and Torfing, "Theoretical Approaches to Metagovernance"; Torfing, "Metagovernance."

6. oxkydo.eth, "Metagovernance in Crypto"; Giove, "Ultimate Guide to Metagovernance."

7. Espejo, "Cybernetics of Governance"; Loeber, "Big Data, Algorithmic Regulation, and the History of the Cybersyn Project in Chile, 1971." On cybernetic governance generally, see Swann, *Anarchist Cybernetics*.

0xkydo.eth. "Metagovernance in Crypto." Substack newsletter. Kydo's Research Log, February 11, 2022. https://kydo.substack.com/p/metagovernance-in-crypto.

Agre, Philip E. *Computation and Human Experience*. Cambridge, UK: Cambridge University Press, 1997.

———. "The Practical Republic: Social Skills and the Progress of Citizenship." In *Community in the Digital Age*, edited by Andrew Feenberg and Darin Barney. Lanham, MD: Rowman and Littlefield, 2004. https://pages.gseis.ucla.edu/faculty/agre/republic.html.

———. "The Structures of Everyday Life." Working Paper. MIT Artificial Intelligence Laboratory, February 1985. https://dspace.mit.edu/handle/1721.1/41473.

Agre, Philip Edward. "The Dynamic Structure of Everyday Life." PhD thesis, Massachusetts Institute of Technology, 1988. https://dspace.mit.edu/handle/1721.1/14422.

Ahonen, Elias. "Rescuing Crypto Workers from Terrible US Job Conditions: John Paller." *Cointelegraph Magazine*, June 22, 2021. https://cointelegraph.com/magazine/2021/06/22/rescuing-crypto-workers-from-terrible-us-job-conditions-john-paller.

Airbnb. "How We're Giving Hosts a Seat at the Table." *Airbnb* (blog), October 30, 2020. https://www.airbnb.com/resources/hosting-homes/a/how-were-giving-hosts-a-seat-at-the-table-283.

Alaimo, Kara. "How the Facebook Arabic Page 'We Are All Khaled Said' Helped Promote the Egyptian Revolution." *Social Media + Society*, October 8, 2015. https://doi.org/10.1177/2056305115604854.

Alba, Davey. "Mark Zuckerberg Is Sure Acting Like Someone Who Might Run for President." *Wired*, January 5, 2017. https://www.wired.com/2017/01/zucks-sure-acting-like-someone-might-run-president/.

Albergotti, Reed. "He Predicted the Dark Side of the Internet 30 Years Ago. Why Did No One Listen?" *Washington Post*, August 12, 2021. https://www.washingtonpost.com/technology/2021/08/12/philip-agre-ai-disappeared/.

Alexander, Amy C., Christian Welzel, Pippa Norris, Erik Voeten, Roberto Stefan Foa, and Yascha Mounk. "Online Exchange on 'Democratic Deconsolidation.'" *Journal of Democracy*, April 28, 2017. https://www.journalofdemocracy.org/online-exchange-democratic-de consolidation/.

Alexander, Scott. "Meditations on Moloch." *Slate Star Codex* (blog), July 30, 2014. https:// slatestarcodex.com/2014/07/30/meditations-on-moloch/.

Ali, Christopher. *Farm Fresh Broadband: The Politics of Rural Connectivity*. Information Policy. Cambridge, MA: MIT Press, 2021.

Ali, Syed Mustafa. "A Brief Introduction to Decolonial Computing." *XRDS: Crossroads, The ACM Magazine for Students* 22, no. 4 (June 13, 2016): 16–21. https://doi.org /10.1145/2930886.

Alkhatib, Ali. "To Live in Their Utopia: Why Algorithmic Systems Create Absurd Outcomes." In *CHI '21: Proceedings of the 2021 CHI Conference on Human Factors in Computing Systems*, 1–9. New York: Association for Computing Machinery, 2021. https://doi .org/10.1145/3411764.3445740.

Allen, Barbara. *Tocqueville, Covenant, and the Democratic Revolution: Harmonizing Earth with Heaven*. Lanham, MD: Lexington Books, 2005.

Allen, Christopher. "The Path to Self-Sovereign Identity." *Life With Alacrity* (blog), April 26, 2016. http://www.lifewithalacrity.com/2016/04/the-path-to-self-soverereign-identity.html.

Alston, Eric. "Blockchain and the Law: Legality, Law-like Characteristics, and Legal Applications." In *The Economics of Blockchain and Cryptocurrency*, edited by James Caton. Cheltenham, UK: Edward Elgar, 2022. https://doi.org/10.4337/9781800882348.00014.

———. "Constitutions and Blockchains: Competitive Governance of Fundamental Rule Sets." Working paper. Center for Growth and Opportunity, Utah State University, March 21, 2019. https://www.growthopportunity.org/research/working-papers/constitutions-and -blockchains.

———. "Political and Economic Institutional Emergence." SSRN, August 18, 2020. https:// doi.org/10.2139/ssrn.3702733.

Anderson, William C. "Ungovernable: An Interview with Lorenzo Kom'boa Ervin." Black Rose/Rosa Negra Anarchist Federation, September 11, 2020. https://blackrosefed.org /ungovernable-interview-lorenzo-komboa-ervin-anderson/.

Andrias, Kate. "Separations of Wealth: Inequality and the Erosion of Checks and Balances." *University of Pennsylvania Journal of Constitutional Law* 18 (2015–2016): 419–504. https://repository.law.umich.edu/articles/1724/.

Aouidef, Yann, Federico Ast, and Bruno Deffains. "Decentralized Justice: A Comparative Analysis of Blockchain Online Dispute Resolution Projects." *Frontiers in Blockchain* 4 (2021). https://doi.org/10.3389/fbloc.2021.564551.

Apache Software Foundation. "Incubating Issues," November 27, 2019. https://cwiki.apache .org/confluence/display/INCUBATOR/Incubating+Issues.

Applebaum, Anne, and Peter Pomerantsev. "How to Put Out Democracy's Dumpster Fire." *The Atlantic*, March 9, 2021. https://www.theatlantic.com/magazine/archive/2021/04/the -internet-doesnt-have-to-be-awful/618079/.

Aragon. "The Fight for Freedom," March 29, 2018. https://www.youtube.com/watch? v=AqjIWmiAidw.

———. "Introducing aragonOS: Say Hi to Modular and Extendable Organizations." *Aragon* (blog), August 5, 2020. https://blog.aragon.org/introducing-aragonos-say-hi-to-modular -and-extendable-organizations-8555af1076f3/.

Aragón, Pablo, Andreas Kaltenbrunner, Antonio Calleja-López, Andrés Pereira, Arnau Monterde, Xabier E. Barandiaran, and Vicenç Gómez. "Deliberative Platform Design: The Case Study of the Online Discussions in Decidim Barcelona." In *Social Informatics*, edited by Giovanni Luca Ciampaglia, Afra Mashhadi, and Taha Yasseri, 277–87. Lecture Notes in Computer Science. Cham, Switzerland: Springer International Publishing, 2017. https://doi.org/10.1007/978-3-319-67256-4_22.

Arcy, Jacquelyn. "Emotion Work: Considering Gender in Digital Labor." *Feminist Media Studies* 16, no. 2 (March 3, 2016): 365–68. https://doi.org/10.1080/14680777.2016 .1138609.

Arendt, Hannah. *The Human Condition*. 2nd ed. 1958. Reprint, University of Chicago Press, 1998.

Argast, Andi, James Brennan, Nabil Echchaibi, Drew Hornbein, Shamika Klassen, Samira Rajabi, and Nathan Schneider. *Sacred Stacks: The Art of Cyborg Community*. Boulder, CO: Media Economies Design Lab, 2023. https://ipfs.metalabel.xyz/ipfs/QmQv VuzF85JDapSAYd8Ke1H2V8ey5je28hbN1UUJn6Hu6e/.

Aristotle. *Nicomachean Ethics*. Translated by W. D. Ross. Cambridge, MA: Internet Classics Archive, 2009. http://classics.mit.edu/Aristotle/nicomachaen.html.

Arndt, David. *Arendt on the Political*. Cambridge, UK: Cambridge University Press, 2019. https://doi.org/10.1017/9781108653282.

Arriola, Tachi. "The Feminist Radio Collective of Peru: Women . . . On the Air." In *A Passion for Radio: Radio Waves and Community*, edited by Bruce Girard, 6. http://comunica.org /passion/pdf/chapter11.pdf. First published 1992. Montreal: Black Rose Books, 2001.

Association for Progressive Communications. "Feminist Principles of the Internet," 2014. https://feministinternet.net/en/.

Atanassow, Ewa. "Colonization and Democracy: Tocqueville Reconsidered." *American Political Science Review* 111, no. 1 (February 2017): 83–96. https://doi.org/10.1017 /S0003055416000630.

Atzori, Marcella. "Blockchain Technology and Decentralized Governance: Is the State Still Necessary?" SSRN, December 2015. https://dx.doi.org/10.2139/ssrn.2709713.

Avila, Renata. "Against Digital Colonialism." In *Platforming Equality: Policy Challenges for the Digital Economy*. Autonomy, 2020. https://autonomy.work/portfolio/platforming equality/.

Ba, Cheick Tidiane, Matteo Zignani, and Sabrina Gaito. "The Role of Groups in a User Migration across Blockchain-Based Online Social Media." In *2022 IEEE International Conference on Pervasive Computing and Communications Workshops and Other Affiliated Events (PerCom Workshops)*, 291–96, 2022. https://doi.org/10.1109/PerComWorkshops 53856.2022.9767453.

Babbitt, Dave, and Joel Dietz. "Crypto-Economic Design: A Proposed Agent-Based Modeling Effort." In *18th Annual Meeting on Agent-Based Modeling & Simulation*. University of Notre Dame, 2014.

Balkin, Jack M. "How to Regulate (and Not Regulate) Social Media." *Journal of Free Speech Law* 1 (2021–2022): 71–96. https://www.journaloffreespeechlaw.org/balkin.pdf.

Barandiaran, Xabier, Antonio Calleja, Arnau Monterde, Pablo Aragón, Juan Linares, Carol Romero, and Andrés Pereira. "Decidim: Redes políticas y tecnopolíticas para la democracia en red." *Recerca*, no. 21 (2017): 137–50. https://doi.org/10.6035/Recerca.2017.21.8.

Barbrook, Richard, and Andy Cameron. "The Californian Ideology." *Science as Culture* 6, no. 1 (1996): 44–72. https://doi.org/10.1080/09505439609526455.

Barnett, Jeremy, and Philip Treleaven. "Algorithmic Dispute Resolution of Professional Dispute Resolution Using AI and Blockchain Technologies." *The Computer Journal* 61, no. 3 (March 1, 2018): 399–408. https://doi.org/10.1093/comjnl/bxx103.

Bartels, Larry M. *Democracy Erodes from the Top: Leaders, Citizens, and the Challenge of Populism in Europe.* Princeton, NJ: Princeton University Press, 2023.

Bauwens, Michel, Vasilis Kostakis, and Alex Pazaitis. *Peer to Peer: The Commons Manifesto.* London: University of Westminster Press, 2019. https://doi.org/10.16997/book33.

BBS: The Documentary. Bovine Ignition Systems, 2005. http://archive.org/details/BBS.The .Documentary.

Beck, Roman, Christoph Müller-Bloch, and John Leslie King. "Governance in the Blockchain Economy: A Framework and Research Agenda." *Journal of the Association for Information Systems* 19, no. 10 (2018): 1020–34. https://doi.org/10.17705/1jais.00518.

Bednar, Jenna. *The Robust Federation: Principles of Design.* Cambridge, UK: Cambridge University Press, 2009.

Bellah, Robert N. "Civil Religion in America." *Daedalus* 96, no. 1 (1967): 1–21. https://www .jstor.org/stable/20027022.

bendi. "Introducing Flexible Voting: An Extension to the Governor Enabling New Voting Mechanisms." Gitcoin Governance forum, July 14, 2022. https://gov.gitcoin.co/t/intro ducing-flexible-voting-an-extension-to-the-governor-enabling-new-voting-mecha nisms/11115.

Benjamin, Ruha. *Race after Technology: Abolitionist Tools for the New Jim Code.* Medford, MA: Polity, 2019.

Benkler, Yochai. *The Wealth of Networks: How Social Production Transforms Markets and Freedom.* New Haven, CT: Yale University Press, 2006.

Bernal, Carlos. "How Constitutional Crowdsourcing Can Enhance Legitimacy in Constitution Making." In *Comparative Constitution Making,* edited by David Landau and Hanna Lerner, 235–56. Cheltenham, UK: Edward Elgar, 2019. https://doi.org/10.433 7/9781785365263.00017.

Berners-Lee, Tim. *Weaving the Web: The Original Design and Ultimate Destiny of the World Wide Web by Its Inventor.* San Francisco: HarperSanFrancisco, 1999.

Bevensee, Emmi, and Rebellious Data LLC. "The Decentralized Web of Hate." *Rebellious Data LLC* (blog), September 2020. https://rebelliousdata.com/wp-content/uploads/2020 /10/P2P-Hate-Report.pdf.

Bevir, Mark. *Key Concepts in Governance.* Thousand Oaks, CA: SAGE Publications, 2009.

Big-8.org. "Big-8 Usenet Hierarchies," January 20, 2013. https://www.big-8.org/wiki/Big-8 _Usenet_hierarchies.

———. "Moderated Newsgroups," July 19, 2012. https://www.big-8.org/wiki/Moderated _Newsgroups.

#BLM10. "It Is Time for Accountability," November 30, 2020. https://www.blmchapterstate ment.com/no1/.

Bird, Stewart, Rene Lichtman, and Peter Gessner, produced in association with the League of Revolutionary Black Workers. *Finally Got the News.* Icarus Films, 1970. http://icarus films.com/if-fin.

Bogaerts, Geert-Jan, José Van Dijck, and Ethan Zuckerman. "Creating PublicSpaces." *Digital Government: Research and Practice* 4, no. 2 (February 13, 2023): 1–13. https://doi .org/10.1145/3582578.

Boggs, Grace. *Living for Change*. 1998. Reprint, Minneapolis: University of Minnesota Press, 2016.

Boggs, Grace Lee, and Scott Kurashige. *The Next American Revolution: Sustainable Activism for the Twenty-First Century*. Berkeley: University of California Press, 2012.

Boggs, James, and Grace Lee Boggs. *Revolution and Evolution in the Twentieth Century*. New York: Monthly Review Press, 1974.

Boiron, Marc. "Sufficient Decentralization: A Playbook for Web3 Builders and Lawyers." Variant, August 2, 2022. https://variant.fund/articles/sufficient-decentralization/.

Bollier, David, and Silke Helfrich. *Patterns of Commoning*. Amherst, MA: Commons Strategy Group and Off the Common Press, 2015.

Bonneau, Joseph, Soren Preibusch, Jonathan Anderson, Richard Clayton, and Ross Anderson. "Democracy Theatre: Comments on Facebook's Proposed Governance Scheme." University of Cambridge Computer Laboratory, March 29, 2009. https://jbonneau.com/doc/BPACA09-unpublished-facebook_comments.pdf.

Börgers, Tilman. *An Introduction to the Theory of Mechanism Design*. Oxford, UK: Oxford University Press, 2015. https://doi.org/10.1093/acprof:oso/9780199734023.001.0001.

Boswell, John, Rikki Dean, and Graham Smith. "Integrating Citizen Deliberation into Climate Governance: Lessons on Robust Design from Six Climate Assemblies." *Public Administration* 101, no. 1 (August 1, 2022): 182–200. https://doi.org/10.1111/padm.12883.

Bowles, Nellie. "A Dark Consensus about Screens and Kids Begins to Emerge in Silicon Valley." Style. *New York Times*, October 26, 2018. https://www.nytimes.com/2018/10/26/style/phones-children-silicon-valley.html.

Bowles, Samuel. *The Moral Economy: Why Good Incentives Are No Substitute for Good Citizens*. New Haven, CT: Yale University Press, 2016.

Boxell, Levi, and Zachary Steinert-Threlkeld. "Taxing Dissent: The Impact of a Social Media Tax in Uganda." *World Development* 158 (October 1, 2022). https://doi.org/10.1016/j.worlddev.2022.105950.

Braman, Sandra. "The Irony of Internet Governance Research: Metagovernance as Context." In *Researching Internet Governance: Methods, Frameworks, Futures*, edited by Laura DeNardis, Derrick Cogburn, Nanette S. Levinson, and Francesca Musiani. Information Policy. Cambridge, MA: MIT Press, 2020. https://direct.mit.edu/books/oa-monograph/4936/chapter/625906/The-Irony-of-Internet-Governance-Research.

Brandel, Jennifer, Mara Zepeda, Astrid Scholz, and Aniyia Williams. "Sex & Startups." *Zebras Unite* (blog), February 17, 2016. https://medium.com/zebras-unite/sex-startups-53f2f63ded49.

Bratton, Benjamin H. *The Stack: On Software and Sovereignty*. Cambridge, MA: MIT Press, 2016.

Brekke, Jaya Klara, and Wassim Zuhair Alsindi. "Cryptoeconomics." *Internet Policy Review* 10, no. 2 (April 20, 2021). https://policyreview.info/glossary/cryptoeconomics.

Brennan, Jason. *Against Democracy*. Princeton, NJ: Princeton University Press, 2016.

Brennan, Patrick McKinley. "Subsidiarity in the Tradition of Catholic Social Doctrine." In *Global Perspectives on Subsidiarity*, edited by Michelle Evans and Augusto Zimmermann, 29–47. Ius Gentium: Comparative Perspectives on Law and Justice. Dordrecht: Springer Netherlands, 2014. https://doi.org/10.1007/978-94-017-8810-6_3.

Bretherton, Luke. "Democracy, Society and Truth: An Exploration of Catholic Social Teaching." *Scottish Journal of Theology* 69, no. 3 (August 2016): 267–80. https://doi.org/10.1017/S0036930616000284.

Bria, Francesca. "Barcelona Digital City: Putting Technology at the Service of People." Barcelona: Ajuntament de Barcelona, 2019. https://ajuntament.barcelona.cat/digital/sites/default/files/pla_barcelona_digital_city_in.pdf.

brown, adrienne maree. *Emergent Strategy: Shaping Change, Changing Worlds*. Chico, CA: AK Press, 2017. Epub.

——. *Holding Change: The Way of Emergent Strategy Facilitation and Mediation*. Chico, CA: AK Press, 2021. Epub.

——. *Pleasure Activism: The Politics of Feeling Good*. Chico, CA: AK Press, 2019. Epub.

——. "Post Nationalism in the Age of Cooptation and Other Dumpster Fires." Personal blog. January 15, 2021. https://adriennemareebrown.net/2021/01/15/post-nationalism-in-the-age-of-cooptation-and-other-dumpster-fires/.

——. *We Will Not Cancel Us: And Other Dreams of Transformative Justice*. Chico, CA: AK Press, 2020.

——. "What Is/Isn't Transformative Justice?" In *Beyond Survival: Strategies and Stories from the Transformative Justice Movement*, edited by Ejeris Dixon and Leah Lakshmi Piepzna-Samarasinha. Chico, CA: AK Press, 2020. Epub.

brown, adrienne maree, and Walidah Imarisha. *Octavia's Brood: Science Fiction Stories from Social Justice Movements*. Chico, CA: AK Press, 2015. Epub.

Brown, R. M. "Spinning without Touching the Wheel: Anticolonialism, Indian Nationalism, and the Deployment of Symbol." *Comparative Studies of South Asia, Africa and the Middle East* 29, no. 2 (January 1, 2009): 230–45. https://doi.org/10.1215/1089201X-2009-006.

Brown, Wendy. *Undoing the Demos: Neoliberalism's Stealth Revolution*. New York: Zone Books, 2015.

Browne, Simone. *Dark Matters: On the Surveillance of Blackness*. Durham, NC: Duke University Press, 2015.

Brunton, Finn. *Digital Cash: The Unknown History of the Anarchists, Utopians, and Technologists Who Created Cryptocurrency*. Princeton, NJ: Princeton University Press, 2020.

Bucher, Taina. "The Algorithmic Imaginary: Exploring the Ordinary Affects of Facebook Algorithms." *Information, Communication & Society* 20, no. 1 (January 2, 2017): 30–44. https://doi.org/10.1080/1369118X.2016.1154086.

Buolamwini, Joy, and Timnit Gebru. "Gender Shades: Intersectional Accuracy Disparities in Commercial Gender Classification." *Proceedings of the 1st Conference on Fairness, Accountability and Transparency* 81 (2018): 77–91. Proceedings of Machine Learning Research, http://proceedings.mlr.press/v81/buolamwini18a.html.

Burr, Danielle, head of Federal Affairs, Uber Technologies, Inc., to Brent J. Fields, U.S. Securities and Exchange Commission. Memorandum, "Re: Concept Release on Compensatory Securities Offerings and Sales," October 11, 2018. https://www.sec.gov/comments/s7-18-18/s71818-4510185-175992.pdf.

Burrington, Ingrid. *Networks of New York: An Illustrated Field Guide to Urban Internet Infrastructure*. Brooklyn: Melville House, 2016.

Buterin, Vitalik. "Blockchain Voting Is Overrated Among Uninformed People but Underrated Among Informed People." *Vitalik Buterin's Website* (blog), May 25, 2021. https://vitalik.ca/general/2021/05/25/voting2.html.

——. "Governance, Part 2: Plutocracy Is Still Bad." *Vitalik Buterin's Website* (blog), March 28, 2018. https://vitalik.ca/general/2018/03/28/plutocracy.html.

———. "Moving Beyond Coin Voting Governance." *Vitalik Buterin's Website* (blog), August 16, 2021. https://vitalik.ca/general/2021/08/16/voting3.html.

———. "On Nathan Schneider on the Limits of Cryptoeconomics." *Vitalik Buterin's Website* (blog), September 26, 2021. https://vitalik.ca/general/2021/09/26/limits.html.

———. *Proof of Stake: The Making of Ethereum and the Philosophy of Blockchains*. Edited by Nathan Schneider. New York: Seven Stories Press, 2022.

———. "Soulbound." *Vitalik Buterin's Website* (blog), January 26, 2022. https://vitalik.ca/general/2022/01/26/soulbound.html.

Buterin, Vitalik, Zoë Hitzig, and E. Glen Weyl. "Liberal Radicalism: A Flexible Design For Philanthropic Matching Funds." SSRN, December 1, 2018. https://papers.ssrn.com/abstract=3243656.

Butler, Philip, ed. *Critical Black Futures: Speculative Theories and Explorations*. Singapore: Palgrave Macmillan, 2021. https://link.springer.com/book/10.1007/978-981-15-7880-9.

———. "Newhampton: A Future Forward(ified) Black City in the United States." In *Critical Black Futures: Speculative Theories and Explorations*, edited by Philip Butler, 207–30. Singapore: Springer, 2021. https://doi.org/10.1007/978-981-15-7880-9_11.

Cady, Kathryn A. "Flexible Labor." *Feminist Media Studies* 13, no. 3 (July 1, 2013): 395–414. https://doi.org/10.1080/14680777.2012.678876.

Caelin, Derek. "Decentralized Networks vs. the Trolls." In *Fundamental Challenges to Global Peace and Security: The Future of Humanity*, edited by Hoda Mahmoudi, Michael H. Allen, and Kate Seaman, 143–68. Cham: Springer International Publishing, 2022. https://doi.org/10.1007/978-3-030-79072-1_8.

Caines, Andrew J. "How to Run a Mailing List (Was: Please Vote Today [*sic*]." Full Disclosure mailing list archives, June 12, 2003. https://seclists.org/fulldisclosure/2003/Jun/396.

Caliskan, Koray. "Data Money Makers: An Ethnographic Analysis of a Global Cryptocurrency Community." *British Journal of Sociology*, December 27, 2021. https://doi.org/10.1111/1468-4446.12916.

Canella, Gino. "Racialized Surveillance: Activist Media and the Policing of Black Bodies." *Communication, Culture and Critique* 11, no. 3 (September 1, 2018): 378–98. https://doi.org/10.1093/ccc/tcy013.

Carayannis, Elias G., David F. J. Campbell, and Evangelos Grigoroudis. "Democracy and the Environment: How Political Freedom Is Linked with Environmental Sustainability." *Sustainability* 13, no. 10 (January 2021). https://doi.org/10.3390/su13105522.

Carugati, Federica, and Nathan Schneider. "Governance Archaeology: Research as Ancestry." *Daedalus* 152, no. 1 (Winter 2023). https://www.amacad.org/publication/governance-archaeology-research-ancestry.

Casarosa, Federica. "Transnational Collective Actions for Cross-Border Data Protection Violations." *Internet Policy Review* 9, no. 3 (September 16, 2020). https://policyreview.info/articles/analysis/transnational-collective-actions-cross-border-data-protection-violations.

Castellanos, Asís, and Adrienne Pine. "Berta Cáceres in Her Own Words." Toward Freedom, July 29, 2020. http://towardfreedom.org/story/berta-caceres-in-her-own-words/.

Castells, Manuel. *Networks of Outrage and Hope: Social Movements in the Internet Age*. 2nd ed. Cambridge, UK: Polity, 2015.

Castillo, Orlando. "VOTEMGR: A Vote Manager for FidoNet Systems." *FidoNews*, September 2, 1991. http://textfiles.com/bbs/FIDONET/FIDONEWS/fido0835.nws.

Catlow, Ruth, and Penny Rafferty, eds. *Radical Friends: Decentralised Autonomous Organisations & the Arts*. n.p.: Torque Editions, 2022.

Certeau, Michel de. *The Practice of Everyday Life*. Translated by Steven Rendall. Berkeley: University of California Press, 1988.

Chaplin, Charles. *My Autobiography*. New York: Pocket Books, 1966. http://archive.org/details/myautobiographyi00chap.

Cherry, Miriam A. "Beyond Misclassification: The Digital Transformation of Work." *Comparative Labor Law & Policy Journal* 37, no. 3 (2015–2016): 577–602. https://scholarship.law.slu.edu/faculty/5/.

Chesnut, Rob, general counsel, Airbnb, to Brent J. Fields, U.S. Securities and Exchange Commission. Memorandum, "Re: Request for Comment on Concept Release on Compensatory Securities Offerings and Sales," September 21, 2018. https://www.sec.gov/comments/s7-18-18/s71818-4403356-175575.pdf.

Chi, Eunju, and Hyeok Yong Kwon. "The Trust-Eroding Effect of Perceived Inequality: Evidence from East Asian New Democracies." *The Social Science Journal* 53, no. 3 (September 1, 2016): 318–28. https://doi.org/10.1016/j.soscij.2016.02.008.

Christensen, Ward, and Randy Seuss. "Hobbyist Computerized Bulletin Board." *Byte*, November 1978. http://vintagecomputer.net/cisc367/byte%20nov%201978%20computerized%20BBS%20-%20ward%20christensen.pdf.

Chu, Cho-Wen. "Censorship or Protectionism? Reassessing China's Regulation of Internet Industry." *International Journal of Social Science and Humanity* 7, no. 1 (2017): 28–32. http://ijssh.org/vol7/790-MC26.pdf.

Chwalisz, Claudia, ed. *Innovative Citizen Participation and New Democratic Institutions: Catching the Deliberative Wave*. Paris: Organisation for Economic Co-operation and Development, 2020. https://www.oecd-ilibrary.org/governance/innovative-citizen-participation-and-new-democratic-institutions_339306da-en.

———. "A Movement That's Quietly Reshaping Democracy for the Better." *Noema*, May 12, 2022. https://www.noemamag.com/a-movement-thats-quietly-reshaping-democracy-for-the-better.

———. *The People's Verdict: Adding Informed Citizen Voices to Public Decision-Making*. London: Rowman & Littlefield International, 2017.

Ciarcia, Steve. "Turnkey Bulletin-Board System." *Byte*, December 1985, 93–103. https://archive.org/details/byte-magazine-1985-12/page/n65/mode/2up/.

Cila, Nazli, Gabriele Ferri, Martijn de Waal, Inte Gloerich, and Tara Karpinski. "The Blockchain and the Commons: Dilemmas in the Design of Local Platforms." In *CHI '20: Proceedings of the 2020 CHI Conference on Human Factors in Computing Systems*, 1–14. New York: Association for Computing Machinery, 2020. https://doi.org/10.1145/3313831.3376660.

Citron, Danielle Keats, and Benjamin Wittes. "The Internet Will Not Break: Denying Bad Samaritans 230 Immunity." *Fordham Law Review* 86 (2017): 25.

Clark, Meredith D. "DRAG THEM: A Brief Etymology of So-Called 'Cancel Culture.'" *Communication and the Public* 5, no. 3–4 (September 2020): 88–92. https://doi.org/10.1177/2057047320961562.

Clifton, Judith, and Leslie A. Pal. "The Policy Dilemmas of Blockchain." *Policy and Society* 41, no. 3 (May 1, 2022): 321–27. https://doi.org/10.1093/polsoc/puac025.

COALA. "Model Law for Decentralized Autonomous Organizations." Coalition of Automated Legal Applications, 2021. https://coala.global/wp-content/uploads/2021/06/DAO-Model-Law.pdf.

Coate, John. "Cyberspace Innkeeping: Building Online Community," 1998. http://johncoate.com/Building%20Online%20Community.html.

Cohen, Noam. "A 1970s Essay Predicted Silicon Valley's High-Minded Tyranny." *Wired*, November 15, 2018. https://www.wired.com/story/silicon-valley-tyranny-of-structurelessness/.

Coleman, E. Gabriella. *Coding Freedom: The Ethics and Aesthetics of Hacking*. Princeton, NJ: Princeton University Press, 2013. https://gabriellacoleman.org/Coleman-Coding-Freedom.pdf.

Colfer, Lyra J., and Carliss Y. Baldwin. "The Mirroring Hypothesis: Theory, Evidence, and Exceptions." *Industrial and Corporate Change* 25, no. 5 (October 1, 2016): 709–38. https://doi.org/10.1093/icc/dtw027.

Computational Democracy Project. "Polis," January 15, 2016. https://web.archive.org/web/20160115032708/https://pol.is/.

Cong, Wanshu. "Contesting Freedom of Information: Capitalism, Development, and the Third World." *Asian Journal of International Law* 13, no. 1 (July 25, 2022): 46–75. https://doi.org/10.1017/S2044251322000467.

Conger, Kate. "'Master,' 'Slave' and the Fight over Offensive Terms in Computing." Technology. *New York Times*, April 13, 2021. https://www.nytimes.com/2021/04/13/technology/racist-computer-engineering-terms-ietf.html.

Conte de Leon, Daniel, Antonius Q. Stalick, Ananth A. Jillepalli, Michael A. Haney, and Frederick T. Sheldon. "Blockchain: Properties and Misconceptions." *Asia Pacific Journal of Innovation and Entrepreneurship* 11, no. 3 (January 1, 2017): 286–300. https://doi.org/10.1108/APJIE-12-2017-034.

Conway, Melvin E. "How Do Committees Invent?" *Datamation*, April 1968. http://www.melconway.com/Home/Committees_Paper.html.

Cooper, Frederick. *Colonialism in Question: Theory, Knowledge, History*. Berkeley: University of California Press, 2005.

Cordes, Ashley. "Storying Indigenous Cryptocurrency: Reckoning with the Ghosts of US Settler Colonialism in the Cultural Economy." *Journal of Cultural Economy*, published online August 23, 2022, 1–18. https://doi.org/10.1080/17530350.2022.2110924.

Cosentino, Valerio, Javier Luis Canovas Izquierdo, and Jordi Cabot. "Three Metrics to Explore the Openness of GitHub Projects." *arXiv preprint*, September 15, 2014. http://arxiv.org/abs/1409.4253.

Costanza-Chock, Sasha. *Design Justice: Community-Led Practices to Build the Worlds We Need*. Cambridge, MA: MIT Press, 2020.

Couldry, Nick, and Andreas Hepp. *The Mediated Construction of Reality*. Cambridge, UK: Polity Press, 2016.

Couldry, Nick, and Ulises A. Mejias. "Data Colonialism: Rethinking Big Data's Relation to the Contemporary Subject." *Television & New Media* 20, no. 4 (May 1, 2019): 336–49. https://doi.org/10.1177/1527476418796632.

———. "Making Data Colonialism Liveable: How Might Data's Social Order Be Regulated?" *Internet Policy Review* 8, no. 2 (June 30, 2019). https://doi.org/10.14763/2019.2.1411.

Coulthard, Glen and Matt Hern. "How Flags Divide Us." *Noema*, October 6, 2022. https://www.noemamag.com/capture-the-flag.

Coulthard, Glen Sean. *Red Skin, White Masks: Rejecting the Colonial Politics of Recognition.* Minneapolis: University of Minnesota Press, 2014.

Cowen, Tyler. "Audrey Tang on the Technology of Democracy." Conversations with Tyler, October 7, 2020. https://medium.com/conversations-with-tyler/audrey-tang-tyler-cowen-taiwan-tech-2ddd75e48bdf.

Creative Interventions. *Creative Interventions Toolkit: A Practical Guide to Stop Interpersonal Violence.* Oakland, CA: Creative Interventions, 2018. https://www.creative-interventions.org/wp-content/uploads/2020/10/CI-Toolkit-Final-ENTIRE-Aug-2020-new-cover.pdf.

Cross, Katherine Alejandra. "Twitter's 'Vox Populi' Is a Lie." *Wired*, January 19, 2023. https://www.wired.com/story/twitter-digital-governance-elon-musk-polls/.

Cullors, Patrisse. *An Abolitionist's Handbook: 12 Steps to Changing Yourself and the World.* New York: St. Martin's Press, 2022.

Curtis, Adam, dir. *All Watched Over by Machines of Loving Grace.* Television series. BBC Two, original release May 23–June 6, 2011.

Dailey, Natasha. "Vitalik Buterin Says He Created Ethereum after His Beloved World of Warcraft Character Was Hobbled by the Developers, Awakening Him to the 'Horrors Centralized Services Can Bring.'" *Markets Insider*, October 5, 2021. https://markets.businessinsider.com/news/currencies/vitalik-buterin-created-ethereum-following-world-of-warcraft-debacle-2021-10.

Davies, William. *The Limits of Neoliberalism: Authority, Sovereignty and the Logic of Competition.* London: SAGE Publications, 2014. https://doi.org/10.4135/9781473906075.

Davis, Angela Y. *Abolition Democracy: Beyond Empire, Prisons, and Torture.* New York: Seven Stories Press, 2005.

Day, Dorothy. *The Long Loneliness: The Autobiography of the Legendary Catholic Social Activist.* 1952. Reprint, New York: HarperOne, 2017.

Dean, Jodi. "Communicative Capitalism: Circulation and the Foreclosure of Politics." *Cultural Politics* 1, no. 1 (March 1, 2005): 51–74. https://doi.org/10.2752/174321905778054845.

De Filippi, Primavera, Morshed Mannan, and Wessel Reijers. "The Alegality of Blockchain Technology." *Policy and Society* 41, no. 3 (May 1, 2022): 358–72. https://doi.org/10.1093/polsoc/puac006.

———. "Blockchain as a Confidence Machine: The Problem of Trust and Challenges of Governance." *Technology in Society* 62 (2020). https://doi.org/10.1016/j.techsoc.2020.101284.

De Filippi, Primavera, and Aaron Wright. *Blockchain and the Law: The Rule of Code.* Cambridge, MA: Harvard University Press, 2018.

Desai, Mahadev. "Charlie Chaplin Meets Gandhiji." *Young India*, October 8, 1931. Reprinted in Nachiketa Desai, "Gandhiji Inspired Charlie Chaplin to Make His Classic Movie 'Modern Times.'" *The Leaflet*, October 10, 2019. https://www.theleaflet.in/gandhiji-inspired-charlie-chaplin-to-make-his-classic-movie-modern-times/.

Detroit Digital Justice Coalition. *How to DiscoTech.* 4. Detroit, MI: Allied Media Projects, 2012. https://detroitcommunitytech.org/system/tdf/librarypdfs/how-to-discotech.pdf.

Devere, Heather. "The Academic Debate on Friendship and Politics." *AMITY: The Journal of Friendship Studies* 1, no. 1 (2013): 5–32. https://doi.org/10.5518/AMITY/2.

Devere, Heather, and Graham M. Smith. "Friendship and Politics." *Political Studies Review* 8, no. 3 (September 2010): 341–56. https://doi.org/10.1111/j.1478-9302.2010.00214.x.

Diamond, Larry. "Democracy's Arc: From Resurgent to Imperiled." *Journal of Democracy* 33, no. 1 (January 2022). https://journalofdemocracy.org/articles/democracys-arc-from-resurgent-to-imperiled/.

Dibbell, Julian. "A Rape in Cyberspace." *The Village Voice*, December 23, 1993. http://www.juliandibbell.com/texts/bungle_vv.html.

Dickie, Bonnie, dir. *Hollow Water*, 2000. https://www.nfb.ca/film/hollow_water/.

Dillard, Cynthia B. "To Experience Joy: Musings on Endarkened Feminisms, Friendship, and Scholarship." *International Journal of Qualitative Studies in Education* 32, no. 2 (February 7, 2019): 112–17. https://doi.org/10.1080/09518398.2018.1533149.

Dinan, Matthew D. "Keeping the Old Name: Derrida and the Deconstructive Foundations of Democracy." *European Journal of Political Theory* 13, no. 1 (January 1, 2014): 61–77. https://doi.org/10.1177/1474885112474441.

Dixon, Ejeris, and Leah Lakshmi Piepzna-Samarasinha, eds. *Beyond Survival: Strategies and Stories from the Transformative Justice Movement*. Chico, CA: AK Press, 2020. Epub.

djahnie. "How Does the Conflict Resolution Feature Work?" Foodsaving Worldwide, March 15, 2019. https://community.foodsaving.world/t/info-how-does-the-conflict-resolution-feature-work/254.

Does, Ramon van der, and Vincent Jacquet. "Small-Scale Deliberation and Mass Democracy: A Systematic Review of the Spillover Effects of Deliberative Minipublics." *Political Studies*, May 5, 2021. https://doi.org/10.1177/00323217211007278.

Donovan, Joan, Emily Dreyfuss, and Brian Friedberg. *Meme Wars: The Untold Story of the Online Battles Upending Democracy in America*. New York: Bloomsbury Publishing, 2022.

Double Union. "Base Assumptions." n.d., accessed November 24, 2021. https://doubleunion.org/base_assumptions/.

Douek, Evelyn. "Facebook's Oversight Board: Move Fast with Stable Infrastructure and Humility." *North Carolina Journal of Law & Technology* 21 (2019–2020): 1. https://scholarship.law.unc.edu/ncjolt/vol21/iss1/2/.

Doyle, Jack. *Lines across the Land: Rural Electric Cooperatives, the Changing Politics of Energy in Rural America*. Washington, DC: The Rural Land & Energy Project, Environmental Policy Institute, 1979.

Driscoll, Kevin. *The Modem World: A Prehistory of Social Media*. New Haven, CT: Yale University Press, 2022.

———. "Thou Shalt Love Thy BBS: A Framework for the Moderation of Online Communities." Paper presented at Computer Networks Histories: Local, National and Transnational Perspectives. Università della Svizzera Italiana, Lugano, Switzerland, 2017. https://ahc-ch.ch/wp-ahc21/wp-content/uploads/21-1-Driscoll.pdf.

Duarte, Marisa Elena, Morgan Vigil-Hayes, Sandra Littletree, and Miranda Belarde-Lewis. "'Of Course, Data Can Never Fully Represent Reality': Assessing the Relationship between 'Indigenous Data' and 'Indigenous Knowledge,' 'Traditional Ecological Knowledge,' and 'Traditional Knowledge.'" *Human Biology* 91, no. 3 (Summer 2019): 163–78. https://doi.org/doi.org/10.13110/humanbiology.91.3.03.

Du Bois, W. E. B. *Black Reconstruction in America*. 1935. Reprint, London: Routledge, 2017.
———. "Of the Ruling of Men." In *Darkwater: Voices from Within the Veil*. New York: Harcourt, Brace, 1920. http://www.webdubois.org/lectures/DuBois;OfTheRulingOfMen.html.
Duffy, Kim Hamilton. "The Time for Self-Sovereign Identity Is Now." *Learning Machine* (blog), November 7, 2017. https://medium.com/learning-machine-blog/the-time-for-self -sovereign-identity-is-now-222aab97041b.
Edge, Jake. "Python Gets a New Governance Model." LWN.net, December 18, 2018. https:// old.lwn.net/Articles/775105/.
Edgerton, David. *The Shock of the Old: Technology and Global History Since 1900*. London: Profile Books, 2006.
Edwards, Paul N., Steven J. Jackson, Geoffrey C. Bowker, and Cory Philip Knobel. "Understanding Infrastructure: Dynamics, Tensions, and Design." Report from History and Theory of Infrastructure: Lessons for New Scientific Cyberinfrastructures workshop, January 2007. http://deepblue.lib.umich.edu/handle/2027.42/49353.
Eglash, Ron. "Broken Metaphor: The Master-Slave Analogy in Technical Literature." *Technology and Culture* 48, no. 2 (2007): 360–69. https://doi.org/10.1353/tech.2007.0066.
Ehmke, Coraline Ada. "Codes of Conduct: When Being Excellent Is Not Enough." *Model View Culture*, December 10, 2014. https://modelviewculture.com/pieces/codes-of-conduct -when-being-excellent-is-not-enough.
———. "A Six-Month Retrospective on Ethical Open Source." *Model View Culture*, April 16, 2020. https://modelviewculture.com/pieces/a-six-month-retrospective-on-ethical-open -source.
El Faqir, Youssef, Javier Arroyo, and Samer Hassan. "An Overview of Decentralized Autonomous Organizations on the Blockchain." In *Proceedings of the 16th International Symposium on Open Collaboration*, Association for Computing Machinery, August 2020, 1–8. https://doi.org/10.1145/3412569.3412579.
Emmett, Jeff. "Conviction Voting: A Novel Continuous Decision Making Alternative to Governance." *Commons Stack* (blog), November 18, 2019. https://medium.com/commonsstack /conviction-voting-a-novel-continuous-decision-making-alternative-to-governance -62e215ad2b3d.
Escobar, Arturo. *Designs for the Pluriverse: Radical Interdependence, Autonomy, and the Making of Worlds*. Durham, NC: Duke University Press, 2018.
Espejo, Raul. "Cybernetics of Governance: The Cybersyn Project 1971." In *Social Systems and Design*, edited by Gary S. Metcalf, 71–90. Translational Systems Sciences. Tokyo: Springer Japan, 2014. https://doi.org/10.1007/978-4-431-54478-4_3.
Eth_man. "Everything about 1Hive in One Place—newFAQ in Development." 1Hive forum, September 29, 2020. https://forum.1hive.org/t/everything-about-1hive-in-one-place-new faq-in-development/180.
Eubanks, Virginia. *Automating Inequality: How High-Tech Tools Profile, Police, and Punish the Poor*. New York: St. Martin's Press, 2018.
Evans, Sandra K., Katy E. Pearce, Jessica Vitak, and Jeffrey W. Treem. "Explicating Affordances: A Conceptual Framework for Understanding Affordances in Communication Research." *Journal of Computer-Mediated Communication* 22, no. 1 (January 1, 2017): 35–52. https://doi.org/10.1111/jcc4.12180.
Fan, Jenny, and Amy X. Zhang. "Digital Juries: A Civics-Oriented Approach to Platform Governance." In *CHI '20: Proceedings of the 2020 CHI Conference on Human Factors in*

Computing Systems, Honolulu, April 25–30, 1–14. New York: Association for Computing Machinery, 2020. http://dx.doi.org/10.1145/3313831.3376293.

Fanon, Frantz. *The Wretched of the Earth*. 1961. Reprint, New York: Grove Weidenfeld, 1963.

Fansher, Madison, Shruthi Sai Chivukula, and Colin M. Gray. "#Darkpatterns: UX Practitioner Conversations about Ethical Design." In *Extended Abstracts of the 2018 CHI Conference on Human Factors in Computing Systems—CHI '18*, 1–6. Montreal: ACM Press, 2018. https://doi.org/10.1145/3170427.3188553.

Federici, Silvia. *Caliban and the Witch: Women, the Body and Primitive Accumulation*. Brooklyn: Autonomedia, 2004.

———. *Revolution at Point Zero: Housework, Reproduction, and Feminist Struggle*. Oakland, CA: PM Press, 2012.

Federman, Mark. "The Penguinist Discourse: A Critical Application of Open Source Software Project Management to Organization Development." *Organization Development Journal* 24, no. 2 (2006): 89–100.

Feiner, Lauren, and Salvador Rodriguez. "Mark Zuckerberg: Facebook Spends More on Safety Than Twitter's Whole Revenue for the Year." Technology, CNBC, May 23, 2019. https://www.cnbc.com/2019/05/23/facebook-fake-account-takedowns-doubled-q4-2018-vs-q1-2019.html.

Felski, Rita. "The Invention of Everyday Life." *New Formation*, no. 39 (Winter 1999): 15–31.

Ferraro, Fabrizio, Jeffrey Pfeffer, and Robert I. Sutton. "Economics Language and Assumptions: How Theories Can Become Self-Fulfilling." *The Academy of Management Review* 30, no. 1 (2005): 8–24. https://doi.org/10.2307/20159091.

Ferreira, Daniel, Jin Li, and Radoslawa Nikolowa. "Corporate Capture of Blockchain Governance." Finance Working Paper No. 593/2019. European Corporate Governance Institute, 2019. SSRN, https://www.ssrn.com/abstract=3320437.

Fiesler, Casey, Jialun "Aaron" Jiang, Joshua McCann, Kyle Frye, and Jed R. Brubaker. "Reddit Rules! Characterizing an Ecosystem of Governance." In *Twelfth International AAAI Conference on Web and Social Media*, 2018. https://www.aaai.org/ocs/index.php/ICWSM/ICWSM18/paper/viewPaper/17898.

Fiesler, Casey, Shannon Morrison, and Amy S. Bruckman. "An Archive of Their Own: A Case Study of Feminist HCI and Values in Design." In *CHI '16: Proceedings of the 2016 CHI Conference on Human Factors in Computing Systems*, San Jose, CA, May 7–12, 2574–85. New York: ACM, 2016. https://doi.org/10.1145/2858036.2858409.

Finley, Klint. "The Woman Bringing Civility to Open Source Projects." *Wired*, September 26, 2018. https://www.wired.com/story/woman-bringing-civility-to-open-source-projects/.

Fish, Adam, Luis F. R. Murillo, Lilly Nguyen, Aaron Panofsky, and Christopher M. Kelty. "Birds of the Internet: Towards a Field Guide to the Organization and Governance of Participation." *Journal of Cultural Economy* 4, no. 2 (May 2011): 157–87. https://doi.org/10.1080/17530350.2011.563069.

Fisk, Catherine L. "Hollywood Writers and the Gig Economy." *University of Chicago Legal Forum* 2017 (2017): 177. https://chicagounbound.uchicago.edu/uclf/vol2017/iss1/8/.

Fitzpatrick, John. "'You Never Know When It's Going to Explode.'" *Living Marxism*, April 1989. https://www.marxists.org/archive/james-clr/works/1989/04/interview.html.

Foa, Roberto Stefan, and Yascha Mounk. "The Signs of Deconsolidation." *Journal of Democracy* 28, no. 1 (January 10, 2017): 5–15. https://doi.org/10.1353/jod.2017.0000.

Follesdal, Andreas. "Survey Article: Subsidiarity." *Journal of Political Philosophy* 6, no. 2 (June 1998): 190–218. https://doi.org/10.1111/1467-9760.00052.

Fotos, Michael A. "Vincent Ostrom's Revolutionary Science of Association." *Public Choice* 163, no. 1 (April 1, 2015): 67–83. https://doi.org/10.1007/s11127-015-0235-1.

Frank, Aaron B., Elizabeth M. Bartels, Adam R. Grissom, Jonathan S. Blake, Gabrielle Tarini, Kelly Elizabeth Eusebi, Joseph N. Mait, et al. "Adaptive Engagement for Undergoverned Spaces: Concepts, Challenges, and Prospects for New Approaches." RAND Corporation, July 13, 2022. https://www.rand.org/pubs/research_reports/RRA1275-1.html.

Freeman, Jo. "The Tyranny of Structurelessness." *Berkeley Journal of Sociology* 17 (1972): 151–64. https://www.jstor.org/stable/41035187.

Freuler, Juan Ortiz. "The Case for a Digital Non-Aligned Movement." *openDemocracy*, June 27, 2020. https://www.opendemocracy.net/en/oureconomy/case-digital-non-aligned -movement/.

Frey, Seth, and Nathan Schneider. "Effective Voice: Beyond Exit and Affect in Online Communities." *New Media & Society*, 2021. https://doi.org/10.1177/14614448211044025.

Frey, Seth, and Robert W. Sumner. "Emergence of Integrated Institutions in a Large Population of Self-Governing Communities." *PLOS ONE* 14, no. 7 (July 11, 2019). https://doi.org /10.1371/journal.pone.0216335.

Friedman, Marilyn. "Autonomy and Social Relationships: Rethinking the Feminist Critique." In *Feminists Rethink the Self*, edited by Diana Tietjens Meyers. 1997. Reprint, New York: Routledge, 2018.

———. "Feminism and Modern Friendship: Dislocating the Community." *Ethics* 99, no. 2 (January 1989).

Fritsch, Felix. "The Common Factory: Governance and Incentive Systems of Blockchain-Based Social Networks." Urbino, Italy, 2019.

Fritz, Kurt von. *The Theory of the Mixed Constitution in Antiquity: A Critical Analysis of Polybius' Political Ideas*. New York: Arno Press, 1975.

Fuchs, Christian. "M. N. Roy and the Frankfurt School: Socialist Humanism and the Critical Analysis of Communication, Culture, Technology, Fascism and Nationalism." *tripleC: Communication, Capitalism & Critique. Open Access Journal for a Global Sustainable Information Society* 17, no. 2 (October 9, 2019): 249–86. https://doi.org/10.31269 /triplec.v17i2.1118.

Fung, Archon, and Erik Olin Wright. *Deepening Democracy: Institutional Innovations in Empowered Participatory Governance*. The Real Utopias Project, vol. 4. London: Verso, 2003.

Galloway, Alexander R. *Protocol: How Control Exists after Decentralization*. Cambridge, MA: MIT Press, 2006.

Galperin, Hernan, and François Bar. "The Microtelco Opportunity: Evidence from Latin America." *Information Technologies & International Development* 3, no. 2 (December 1, 2006): 73–86. https://itidjournal.org/index.php/itid/article/view/225.

Gandhi, M. K. *Constructive Programme: Its Meaning and Place*. Ahmedabad, India: Navajivan Publishing House, 1941. https://www.gandhiheritageportal.org/mahatma-gandhi -books/constructive-programme-its-meaning-and-place.

Ganley, Dale, and Cliff Lampe. "The Ties That Bind: Social Network Principles in Online Communities." In "Online Communities and Social Network," edited by Wenjing Duan.

Special issue, *Decision Support Systems*, 47, no. 3 (June 1, 2009): 266–74. https://doi.org /10.1016/j.dss.2009.02.013.

Garg, Vaibhav, and L. Jean Camp. "Gandhigiri in Cyberspace: A Novel Approach to Information Ethics." *ACM SIGCAS Computers and Society* 42, no. 1 (August 2012): 9–20. https://doi.org/10.1145/2422512.2422514.

Gargarella, Roberto. "From 'Democratic Erosion' to 'a Conversation among Equals.'" *Revus*, no. 47 (January 26, 2022). https://doi.org/10.4000/revus.8079.

Garza, Alicia. *The Purpose of Power: How We Come Together When We Fall Apart*. New York: One World, 2020.

Getachew, Adom. *Worldmaking After Empire: The Rise and Fall of Self-Determination*. Princeton, NJ: Princeton University Press, 2019.

Gillespie, Tarleton. "Content Moderation, AI, and the Question of Scale." *Big Data & Society* 7, no. 2 (July 1, 2020). https://doi.org/10.1177/2053951720943234.

———. *Custodians of the Internet: Platforms, Content Moderation, and the Hidden Decisions That Shape Social Media*. New Haven, CT: Yale University Press, 2018.

———. "Governance of and by Platforms." In *The SAGE Handbook of Social Media*, edited by Jean Burgess, Alice E. Marwick, and Thomas Poell. Thousand Oaks, CA: SAGE Publications, 2018.

———. "The Politics of 'Platforms.'" *New Media & Society* 12, no. 3 (May 1, 2010): 347–64. https://doi.org/10.1177/1461444809342738.

Giove, Ben. "Ultimate Guide to Metagovernance." *Bankless* (blog), May 26, 2022. https:// www.bankless.com/ultimate-guide-to-metagovernance.

Gipson, Grace D. "Afrofuturism's Musical Princess Janelle Monae: Psychedelic Soul Message Music Infused with a Sci-Fi Twist." In *Afrofuturism 2.0: The Rise of Astro-Blackness*, edited by Reynaldo Anderson and Charles E. Jones. Lanham, MD: Lexington Books, 2015.

Giraudet, Louis-Gaëtan, Bénédicte Apouey, Hazem Arab, Simon Baeckelandt, Philippe Begout, Nicolas Berghmans, Nathalie Blanc, et al. "'Co-construction' in Deliberative Democracy: Lessons from the French Citizens' Convention for Climate." *Humanities and Social Sciences Communications* 9, no. 207 (May 20, 2022). https://hal-enpc.archives -ouvertes.fr/hal-03119539.

Golumbia, David. *The Politics of Bitcoin: Software as Right-Wing Extremism*. Minneapolis: University of Minnesota Press, 2016.

Gonzalez, Lorena. AB-1319: The Cooperative Economy Act. California Legislature, February 19, 2021. https://leginfo.legislature.ca.gov/faces/billTextClient.xhtml?bill_id=202120220AB1319.

Goodman, J. David. "A Year after 'Defund,' Police Departments Get Their Money Back." *New York Times*, October 10, 2021. https://www.nytimes.com/2021/10/10/us/dallas-police-defund .html.

Goodwyn, Lawrence. *The Populist Moment: A Short History of the Agrarian Revolt in America*. New York: Oxford University Press, 1978.

Gordon Nembhard, Jessica. *Collective Courage: A History of African American Cooperative Economic Thought and Practice*. University Park: Penn State University Press, 2014.

Graeber, David, and Marshall Sahlins. *On Kings*. Chicago: HAU, 2017.

Graeber, David, and David Wengrow. *The Dawn of Everything: A New History of Humanity*. New York: Farrar, Straus and Giroux, 2021.

Graham, Paul. "Startup = Growth," September 2012. http://paulgraham.com/growth.html.

Gramlich, John. "Violent Crime Is a Key Midterm Voting Issue, but What Does the Data Say?" Pew Research Center, October 31, 2022. https://www.pewresearch.org/fact-tank/2022/10/31/violent-crime-is-a-key-midterm-voting-issue-but-what-does-the-data-say/.

Gray, Jeffery. "Transgression, Release and 'Moloch.'" In *The Taboo*, edited by Blake Hobby, 37–49. Bloom's Literary Themes. New York: Blake's Literary Criticism, 2010.

Gray, Mary L. "Negotiating Identities/Queering Desires: Coming Out Online and the Remediation of the Coming-Out Story." *Journal of Computer-Mediated Communication* 14, no. 4 (July 1, 2009): 1162–89. https://doi.org/10.1111/j.1083-6101.2009.01485.x.

Gray, Mary L., and Siddharth Suri. *Ghost Work: How to Stop Silicon Valley from Building a New Global Underclass.* Boston: Houghton Mifflin Harcourt, 2019.

Gritsenko, Daria, and Matthew Wood. "Algorithmic Governance: A Modes of Governance Approach." *Regulation & Governance*, 2020. https://doi.org/10.1111/rego.12367.

Guinier, Lani. *Lift Every Voice: Turning a Civil Rights Setback Into a New Vision of Social Justice.* New York: Simon and Schuster, 2003.

Guinier, Lani, and Gerald Torres. *The Miner's Canary: Enlisting Race, Resisting Power, Transforming Democracy.* Cambridge, MA: Harvard University Press, 2003.

Gurevich, Liena. "Patriarchy? Paternalism? Motherhood Discourses in Trials of Crimes against Children." *Sociological Perspectives* 51, no. 3 (September 2008): 515–39. https://doi.org/10.1525/sop.2008.51.3.515.

Gurri, Martin. *The Revolt of the Public and the Crisis of Authority in the New Millenium.* 2nd ed. San Francisco: Stripe Press, 2018.

Hafner, Katie. "The Epic Saga of The Well." *Wired*, May 1, 1997. https://www.wired.com/1997/05/ff-well/.

Hafner, Katie, and Matthew Lyon. *Where Wizards Stay Up Late: The Origins of the Internet.* New York: Simon & Schuster, 1996.

Haidt, Jonathan. "Why the Past 10 Years of American Life Have Been Uniquely Stupid." *The Atlantic*, April 11, 2022. https://www.theatlantic.com/magazine/archive/2022/05/social-media-democracy-trust-babel/629369/.

Haidt, Jonathan, and Chris Ball. "Social Media and Political Dysfunction: A Collaborative Review." New York University, ongoing, first posted November 2, 2021. https://tinyurl.com/PoliticalDysfunctionReview.

Haldane, J. B. S. "On Being the Right Size." *Harper's Magazine*, March 26, 1926. http://www.phys.ufl.edu/courses/phy3221/spring10/HaldaneRightSize.pdf.

Halfaker, Aaron, R. Stuart Geiger, Jonathan T. Morgan, and John Riedl. "The Rise and Decline of an Open Collaboration System: How Wikipedia's Reaction to Popularity Is Causing Its Decline." *American Behavioral Scientist* 57, no. 5 (May 1, 2013): 664–88. https://doi.org/10.1177/0002764212469365.

Hall, Stuart. "Encoding/Decoding." In *Culture, Media, Language*, edited by Stuart Hall, Dorothy Hobson, Andrew Love, and Paul Willis, 128–38. London: Hutchinson, 1980.

Han, Anna S. "The Facebook IPO's Face-off with Dual Class Stock Structure." *University of Michigan Journal of Law Reform Online* 45 (2012): 50–55. https://heinonline.org/HOL/P?h=hein.journals/caveat45&i=50.

Harari, Yuval Noah. *Sapiens: A Brief History of Humankind.* New York: Harper, 2015.

Haraway, Donna. "A Cyborg Manifesto: Science, Technology, and Socialist-Feminism in the Late Twentieth Century." In *Simians, Cyborgs, and Women: The Reinvention of Nature.* New York: Routledge, 1991.

Harding, Sandra G. *The Science Question in Feminism*. Ithaca, NY: Cornell University Press, 1986.

Hardt, Michael, and Antonio Negri. *Assembly*. New York: Oxford University Press, 2017.

Harney, Stefano, and Fred Moten. *The Undercommons: Fugitive Planning & Black Study*. Wivenhoe: Minor Compositions, 2013.

Harris, Malcolm. *Palo Alto: A History of California, Capitalism, and the World*. New York: Little, Brown and Company, 2023.

Hartman, Saidiya. *Lose Your Mother: A Journey along the Atlantic Slave Route*. New York: Farrar, Straus and Giroux, 2008.

Hasinoff, Amy, Anna D. Gibson, and Niloufar Salehi. "The Promise of Restorative Justice in Addressing Online Harm." Brookings. TechStream, July 27, 2020. https://www.brookings.edu/techstream/the-promise-of-restorative-justice-in-addressing-online-harm/.

Hasinoff, Amy, and Nathan Schneider. "From Scalability to Subsidiarity in Addressing Online Harm." *Social Media + Society* 8, no. 3 (2022). https://doi.org/10.1177/20563051221126041.

Havel, Václav. "The Power of the Powerless," October 1978. International Center on Nonviolent Conflict. https://www.nonviolent-conflict.org/wp-content/uploads/1979/01/the-power-of-the-powerless.pdf.

Hayden, Tom. *The Zapatista Reader*. New York: Thunder's Mouth Press/Nation Books, 2002.

He, Baogang, and Mark E. Warren. "Authoritarian Deliberation: The Deliberative Turn in Chinese Political Development." *Perspectives on Politics* 9, no. 2 (June 2011): 269–89. https://doi.org/10.1017/S1537592711000892.

Hepp, Andreas. *Deep Mediatization*. London: Routledge, 2019. https://doi.org/10.4324/9781351064903.

Hess, Charlotte, and Elinor Ostrom. "A Framework for Analyzing the Knowledge Commons." In *Understanding Knowledge as a Commons: From Theory to Practice*, 41–81. Cambridge, MA: MIT Press, 2007. https://ieeexplore.ieee.org/document/6284192.

Highmore, Ben. *Everyday Life and Cultural Theory: An Introduction*. London: Routledge, 2002.

Hinman, William. "Digital Asset Transactions: When Howey Met Gary (Plastic)." Presented at the Yahoo Finance All Markets Summit: Crypto, San Francisco, June 14, 2018. https://www.sec.gov/news/speech/speech-hinman-061418.

Hirschman, Albert O. *Exit, Voice, and Loyalty: Responses to Decline in Firms, Organizations, and States*. Cambridge, MA: Harvard University Press, 1970.

Hoback, Cullen, dir. *Q: Into the Storm*. HBO, 2021. https://www.hbo.com/q-into-the-storm.

Hodgson, Matthew. "Matrix and Riot Confirmed as the Basis for France's Secure Instant Messenger App." *Matrix* (blog), April 26, 2018. https://matrix.org/blog/2018/04/26/matrix-and-riot-confirmed-as-the-basis-for-frances-secure-instant-messenger-app.

Holyoake, George Jacob. *The History of Co-Operation*. London: T. F. Unwin, 1908. http://archive.org/details/cu31924002593816.

Hon, Adrian. "How Game Design Principles Can Enhance Democracy." *Noema*, September 20, 2022. https://www.noemamag.com/how-game-design-principles-can-enhance-democracy.

hooks, bell. "Homeplace (A Site of Resistance)." In *Yearning: Race, Gender, and Cultural Politics*, 41–49. Boston: South End Press, 1990.

Hsiao, Yu-Tang, Shu-Yang Lin, Audrey Tang, Darshana Narayanan, and Claudina Sarahe. "vTaiwan: An Empirical Study of Open Consultation Process in Taiwan." *SocArXiv*, July 3, 2018. https://doi.org/10.31235/osf.io/xyhft.

Huhtamo, Erkki, and Jussi Parikka. *Media Archaeology: Approaches, Applications, and Implications*. Berkeley: University of California Press, 2011.

Huizinga, Johan. *Homo Ludens: A Study of the Play-Element in Culture*. 1944. Reprint, London: Routledge, 1949.

Hwang, Sohyeon, and Aaron Shaw. "Rules and Rule-Making in the Five Largest Wikipedias." In *Proceedings of the Sixteenth International AAAI Conference on Web and Social Media*, 11, 2022.

Hyman, Avi. "Twenty Years of ListServ as an Academic Tool." *The Internet and Higher Education* 6, no. 1 (January 1, 2003): 17–24. https://doi.org/10.1016/S1096-7516(02)00159-8.

Illich, Ivan. *Shadow Work*. Boston: Marion Boyars, 1981.

———. *Tools for Conviviality*. New York: Harper & Row, 1973.

Ingersoll, Ernest. "Wampum and Its History." *The American Naturalist* 17, no. 5 (May 1883): 467–79. https://doi.org/10.1086/273355.

Interrupting Criminalization, and Project Nia. One Million Experiments. https://millionexperiments.com/.

Invisible Committee. *The Coming Insurrection*. Intervention 1. Los Angeles: Semiotext(e), 2009.

Irani, Lilly. "Difference and Dependence among Digital Workers: The Case of Amazon Mechanical Turk." *South Atlantic Quarterly* 114, no. 1 (January 1, 2015): 225–34. https://doi.org/10.1215/00382876-2831665.

Irani, Lilly, Janet Vertesi, Paul Dourish, Kavita Philip, and Rebecca E. Grinter. "Postcolonial Computing: A Lens on Design and Development." In *CHI '10: Proceedings of the SIGCHI Conference on Human Factors in Computing Systems*, 1311–20. New York: Association for Computing Machinery, 2010. https://doi.org/10.1145/1753326.1753522.

Iyer, Neema, Garnett Achieng, Favour Borokini, and Uri Ludger. "Automated Imperialism, Expansionist Dreams: Exploring Digital Extractivism in Africa." Pollicy, June 2021. https://pollicy.org/digitalextractivism/.

Jackson, Sarah J., and Brooke Foucault Welles. "Hijacking #MYNYPD: Social Media Dissent and Networked Counterpublics." *Journal of Communication* 65, no. 6 (December 1, 2015): 932–52. https://doi.org/10.1111/jcom.12185.

Jacobs, Renee. "Iroquois Great Law of Peace and the United States Constitution: How the Founding Fathers Ignored the Clan Mothers." *American Indian Law Review* 16 (1991): 497. https://heinonline.org/HOL/Page?handle=hein.journals/aind16&id=503&div=&collection=.

Jäger, Simon, Shakked Noy, and Benjamin Schoefer. "What Does Codetermination Do?" *ILR Review*, December 29, 2021. https://doi.org/10.1177/00197939211065727.

James, C. L. R. *The Black Jacobins: Toussaint L'Ouverture and the San Domingo Revolution*. 1938. Reprint, New York: Vintage Books, 1989. http://archive.org/details/blackjacobinsooclrj.

———. "Every Cook Can Govern." *Correspondence* 2, no. 12 (June 1956). https://www.marxists.org/archive/james-clr/works/1956/06/every-cook.htm.

James, C. L. R., Grace C. Lee, and Pierre Chaulieu. *Facing Reality*. 1958. Reprint, Detroit: Bewick Editions, 1974. https://libcom.org/files/James%20-%20Facing%20Reality.pdf.

Jandrić, Petar, and Ana Kuzmanić. "Digital Postcolonialism." *IADIS International Journal on WWW/Internet* 13, no. 2 (2016): 34–51.

Jarrett, Kylie. "The Relevance of 'Women's Work': Social Reproduction and Immaterial Labor in Digital Media." *Television & New Media* 15, no. 1 (January 1, 2014): 14–29. https://doi.org/10.1177/1527476413487607.

Jay-Z and Kanye West. "No Church in the Wild." In *Watch the Throne*. Def Jam, 2011.

Jemielniak, Dariusz. *Common Knowledge?: An Ethnography of Wikipedia*. Redwood City, CA: Stanford University Press, 2014.

Jessop, Bob. "Governance and Metagovernance: On Reflexivity, Requisite Variety, and Requisite Irony." Department of Sociology, Lancaster University, 2002. https://www .lancaster.ac.uk/fass/resources/sociology-online-papers/papers/jessop-governance -and-metagovernance.pdf.

———. "The Rise of Governance and the Risks of Failure: The Case of Economic Development." *International Social Science Journal* 50, no. 155 (March 1998): 29–45. https:// eprints.lancs.ac.uk/id/eprint/239/.

Jhaver, Shagun, Iris Birman, Eric Gilbert, and Amy Bruckman. "Human-Machine Collaboration for Content Regulation: The Case of Reddit AutoModerator." *ACM Transactions on Computer-Human Interaction* 26, no. 5 (July 19, 2019): 1–35. https://doi.org /10.1145/3338243.

Jhaver, Shagun, Seth Frey, and Amy Zhang. "Designing for Multiple Centers of Power: A Taxonomy of Multi-level Governance in Online Social Platforms," August 27, 2021. http://arxiv.org/abs/2108.12529.

Jin, Dal Yong. "The Construction of Platform Imperialism in the Globalization Era." *tripleC: Communication, Capitalism & Critique* 11, no. 1 (January 11, 2013): 145–72. https://doi.org /10.31269/triplec.v11i1.458.

Johansson, Magnus, Harko Verhagen, and Yubo Kou. "I Am Being Watched by the Tribunal: Trust and Control in Multiplayer Online Battle Arena Games." In *Proceedings of the 10th International Conference on Foundations of Digital Games*, June 22–25, 2015, Pacific Grove, CA, 2015. https://wtf.tw/ref/johansson_2015.pdf.

Johnson, Carolina, H. Jacob Carlson, and Sonya Reynolds. "Testing the Participation Hypothesis: Evidence from Participatory Budgeting." *Political Behavior*, February 25, 2021. https://doi.org/10.1007/s11109-021-09679-w.

Johnson, Eliana, and Eli Stokols. "What Steve Bannon Wants You to Read." *Politico*, February 7, 2017. https://politi.co/2zmbKEZ.

Johnson, Jason. "Why Violent Crime Surged after Police across America Retreated." *USA Today*, April 9, 2021. https://www.usatoday.com/story/opinion/policing/2021/04/09/violent -crime-surged-across-america-after-police-retreated-column/7137565002/.

Johnson, J. R. "They Showed the Way to Labor Emancipation!" *Labor Action*, March 18, 1946. https://www.marxists.org/archive/james-clr/works/1946/03/paris-commune.htm.

Johnston, Hannah, and Chris Land-Kazlauskas. "Organizing On-Demand: Representation, Voice, and Collective Bargaining in the Gig Economy." Geneva: International Labour Organization, 2018. https://www.ilo.org/travail/info/publications/WCMS_624286/lang--en /index.htm.

Jones, Rhett. "Linux Founder Takes Some Time Off to Learn How to Stop Being an Asshole." *Gizmodo*, September 17, 2018. https://gizmodo.com/linux-founder-takes-some-time -off-to-learn-how-to-stop-1829105667.

Kaba, Mariame. "Be Humble." In *Beyond Survival: Strategies and Stories from the Transformative Justice Movement*, edited by Ejeris Dixon and Leah Lakshmi Piepzna-Samarasinha. Chico, CA: AK Press, 2020. Epub.

———. "Yes, We Mean Literally Abolish the Police." Opinion. *New York Times*, June 12, 2020. https://www.nytimes.com/2020/06/12/opinion/sunday/floyd-abolish-defund-police .html.

Kaba, Mariame, and Shira Hassan. *Fumbling towards Repair: A Workbook for Community Accountability Facilitators.* Chicago: Project NIA, 2019.

Kaba, Mariame, and Naomi Murakawa. *We Do This 'til We Free Us: Abolitionist Organizing and Transforming Justice.* Edited by Tamara K. Nopper. Chicago: Haymarket Books, 2021.

Kaba, Mariame, and Andrea Ritchie. *No More Police: A Case for Abolition.* New York: The New Press, 2022.

Karjalainen, Risto. "Governance in Decentralized Networks," 2020. SSRN, https://doi.org /http://dx.doi.org/10.2139/ssrn.3551099.

Katsaros, Matthew, Tom Tyler, Jisu Kim, and Tracey Meares. "Procedural Justice and Self-Governance on Twitter: Unpacking the Experience of Rule Breaking on Twitter." *Journal of Online Trust and Safety* 1, no. 3 (August 31, 2022). https://doi.org/10.54501/jots.v1i3.38.

Kazan, Elia, dir. *On the Waterfront.* Budd Schulberg, screenplay. Horizon Pictures, 1954.

Kelty, Christopher M. *The Participant.* Chicago: University of Chicago Press, 2019.

———. *Two Bits: The Cultural Significance of Free Software.* Experimental Futures. Durham, NC: Duke University Press, 2008.

Kember, Sarah, and Joanna Zylinska. *Life after New Media: Mediation as a Vital Process.* Cambridge, MA: MIT Press, 2012.

King, Patrick. "Introduction to Boggs." *E-Flux,* February 2017. https://www.e-flux.com /journal/79/94671/introduction-to-boggs/.

Kirby, Jay, and Lori Emerson. "As If, or, Using Media Archaeology to Reimagine Past, Present, and Future: An Interview with Lori Emerson." *International Journal of Communication* 10 (June 24, 2016). https://ijoc.org/index.php/ijoc/article/view/4764.

Kirkpatrick, David D. "Who Is Behind QAnon? Linguistic Detectives Find Fingerprints." *New York Times: Technology,* February 19, 2022. https://www.nytimes.com/2022/02/19 /technology/qanon-messages-authors.html.

Klein, Steven. "'Fit to Enter the World': Hannah Arendt on Politics, Economics, and the Welfare State." *American Political Science Review* 108, no. 4 (November 2014): 856–69. https://doi.org/10.1017/S0003055414000409.

Klonick, Kate. "The New Governors: The People, Rules, and Processes Governing Online Speech." *Harvard Law Review* 131 (2018): 1598–1670. https://harvardlawreview.org/print /vol-131/the-new-governors-the-people-rules-and-processes-governing-online -speech/.

Knight, Kyle W., Juliet B. Schor, and Andrew K. Jorgenson. "Wealth Inequality and Carbon Emissions in High-Income Countries." *Social Currents* 4, no. 5 (October 1, 2017): 403–12. https://doi.org/10.1177/2329496517704872.

Kollock, Peter, and Marc Smith. "Managing the Virtual Commons: Cooperation and Conflict in Computer Communities." In *Computer-Mediated Communication: Linguistic, Social, and Cross-Cultural Perspectives,* edited by Susan C. Herring, 109–28. Pragmatics and Beyond New Series, 39. Amsterdam: John Benjamins Publishing Company, 1996. https://doi.org/10.1075/pbns.39.10kol.

Komporozos-Athanasiou, Aris. *Speculative Communities: Living with Uncertainty in a Financialized World.* Chicago: University of Chicago Press, 2022.

König, Thomas, Enric Duran, Niklas Fessler, and Roland Alton. "The Proof-of-Cooperation Blockchain FairCoin," July 2018. https://fair-coin.org/sites/default/files/FairCoin2 _whitepaper_V1.2.pdf.

Kooiman, J., ed. *Modern Governance: New Government-Society Interactions*. London; Newbury Park, CA: Sage, 1993.

Korpas, Lucia M., Seth Frey, and Joshua Tan. "Political, Economic, and Governance Attitudes of Blockchain Users." *Frontiers in Blockchain* 6 (2023). https://www.frontiersin.org/articles/10.3389/fbloc.2023.1125088.

Kosseff, Jeff. *The Twenty-Six Words That Created the Internet*. Ithaca, NY: Cornell University Press, 2019. https://doi.org/10.7591/9781501735783.

Kostakis, Vasilis. "Peer Governance and Wikipedia: Identifying and Understanding the Problems of Wikipedia's Governance." *First Monday* 15, no. 3 (March 12, 2010). https://doi.org/10.5210/fm.v15i3.2613.

Kotliar, Dan M. "Data Orientalism: On the Algorithmic Construction of the Non-Western Other." *Theory and Society*, August 27, 2020. https://doi.org/10.1007/s11186-020-09404-2.

Kou, Yubo, Xinning Gui, Shaozeng Zhang, and Bonnie Nardi. "Managing Disruptive Behavior through Non-Hierarchical Governance: Crowdsourcing in League of Legends and Weibo." *Proceedings of the ACM on Human-Computer Interaction* 1 (December 6, 2017): 1–17. https://doi.org/10.1145/3134697.

Kraidy, Marwan M. "Fun against Fear in the Caliphate: Islamic State's Spectacle and Counter-Spectacle." *Critical Studies in Media Communication* 35, no. 1 (January 1, 2018): 40–56. https://doi.org/10.1080/15295036.2017.1394583.

Kreutler, Kei. "A Prehistory of DAOs." *Gnosis Guild* (blog), July 21, 2021. https://gnosisguild.mirror.xyz/t4F5rItMw4-mlpLZf5JQhElbDfQ2JRVKAzEpanyxW1Q.

———. "Zodiac: The Expansion Pack for DAOs." *Gnosis Guild* (blog), September 14, 2021. https://gnosisguild.mirror.xyz/OuhG5s2X5uSVBx1EK4tKPhnUc91Wh9YMofwSnC8UNcg.

Kropotkin, Peter. *Mutual Aid: A Factor of Evolution*. 1902. Reprint, Standard Ebooks, 2022. https://standardebooks.org/ebooks/peter-kropotkin/mutual-aid.

Kwet, Michael. "Digital Colonialism: US Empire and the New Imperialism in the Global South." *Race & Class* 60, no. 4 (2019): 3–26.

Laffan, Liz. "A New Way of Measuring Openness: The Open Governance Index." *Technology Innovation Management Review* 2, no. 1 (January 2012): 18–24. https://timreview.ca/article/512.

Lampe, Cliff, and Paul Resnick. "Slash(dot) and Burn: Distributed Moderation in a Large Online Conversation Space." In *CHI '04: Proceedings of the SIGCHI Conference on Human Factors in Computing Systems*, 543–50. New York: Association for Computing Machinery, 2004. https://doi.org/10.1145/985692.985761.

Landemore, Hélène. *Open Democracy*. Princeton, NJ: Princeton University Press, 2020.

Łapniewska, Zofia. "Reading Elinor Ostrom through a Gender Perspective." *Feminist Economics* 22, no. 4 (October 1, 2016): 129–51. https://doi.org/10.1080/13545701.2016.1171376.

Leavitt, Alex, and John J. Robinson. "The Role of Information Visibility in Network Gatekeeping: Information Aggregation on Reddit during Crisis Events." In *Proceedings of the 2017 ACM Conference on Computer Supported Cooperative Work and Social Computing*, 1246–61. CSCW '17. New York: ACM, 2017. https://doi.org/10.1145/2998181.2998299.

Lebrón Ortiz, Pedro. "Resisting (Meta) Physical Catastrophes through Acts of Marronage." *Radical Philosophy Review* 23, no. 1 (2020): 35–57. https://doi.org/10.5840/radphilrev202021910.

Lefebvre, Henri. *Critique of Everyday Life*. Translated by John Moore. London: Verso, 1991.

Lemley, Mark A. "The Splinternet." Stanford Law and Economics, Olin Working Paper No. 555. September 28, 2020. SSRN, https://doi.org/10.2139/ssrn.3664027.

Lenin, Vladimir Ilyich. "What Is To Be Done?" In *Collected Works*, 5:347–530. 1902. Reprint, Moscow: Foreign Languages Publishing House, 1961. https://www.marxists.org/archive /lenin/works/1901/witbd/.

Leonard, Andrew. "How Taiwan's Unlikely Digital Minister Hacked the Pandemic." *Wired*, July 23, 2020. https://www.wired.com/story/how-taiwans-unlikely-digital-minister-hacked -the-pandemic/.

Lessig, Lawrence. *Code*. Version 2.0. New York: Basic Books, 2006.

Levi-Faur, David. *The Oxford Handbook of Governance*. Oxford, UK: Oxford University Press, 2012. https://doi.org/10.1093/oxfordhb/9780199560530.001.0001.

Lewis, Ruth, Michael Rowe, and Clare Wiper. "Online Abuse of Feminists as an Emerging Form of Violence against Women and Girls." *The British Journal of Criminology* 57, no. 6 (November 1, 2017): 1462–81. https://doi.org/10.1093/bjc/azw073.

Li, Tania Murray. "Governmentality." *Anthropologica* 49, no. 2 (2007): 275–81. https://www .jstor.org/stable/25605363.

Limerick, Patricia Nelson. *The Legacy of Conquest: The Unbroken Past of the American West*. New York: Norton, 1988.

Lippmann, Walter. "The Basic Problem of Democracy." *Atlantic Monthly*, November 1, 1919. https://www.theatlantic.com/magazine/archive/1919/11/the-basic-problem-of-democ racy/569095/.

———. *Public Opinion*. New York: Harcourt, Brace, 1922.

Loeber, Katharina. "Big Data, Algorithmic Regulation, and the History of the Cybersyn Project in Chile, 1971." In "Big Data and the Human and Social Sciences." Special issue, *Social Sciences* 7, no. 4, 4 (April 2018): 65. https://doi.org/10.3390/socsci7040065.

Lofton, Kathryn. *Consuming Religion*. Chicago: University of Chicago Press, 2017.

Londoño, Wilhelm. "Indigenous Archaeology, Community Archaeology, and Decolonial Archaeology: What Are We Talking About? A Look at the Current Archaeological Theory in South America with Examples." *Archaeologies*, August 14, 2021. https://doi .org/10.1007/s11759-021-09433-y.

Lopez, Alfredo, Jamie McClelland, Eric Goldhagen, Daniel Kahn Gillmor, and Amanda B. Hickman. *The Organic Internet: Organizing History's Largest Social Movement*. Entremundos Publications, 2007. https://mayfirst.coop/files/organicinternet.1.5.pdf.

Lorenz-Spreen, Philipp, Lisa Oswald, Stephan Lewandowsky, and Ralph Hertwig. "A Systematic Review of Worldwide Causal and Correlational Evidence on Digital Media and Democracy." *Nature Human Behaviour*, November 7, 2022. https://doi.org/10.1038 /s41562-022-01460-1.

Lozano-Paredes, Luis H. "Emergent Transportation 'Platforms' in Latin America: Online Communities and Their Governance Models." *Frontiers in Human Dynamics* 3 (2021). https://doi.org/10.3389/fhumd.2021.628556.

Ludlow, Peter, ed. *Crypto Anarchy, Cyberstates, and Pirate Utopias*. Digital Communication. Cambridge, MA: MIT Press, 2001.

Luo, Zhifan, and Muyang Lu. "Participatory Censorship: How Online Fandom Community Facilitates Authoritarian Rule." *New Media & Society*, 2022. https://doi.org/10.1177 /14614448221113923.

Luxemburg, Rosa. "Organizational Questions of Russian Social Democracy." *Neue Zeit*, 1904. https://rosaluxemburg.org/en/material/2741/.

Lynch, Casey R. "Contesting Digital Futures: Urban Politics, Alternative Economies, and the Movement for Technological Sovereignty in Barcelona." *Antipode* 52, no. 3 (2020): 660–80. https://doi.org/10.1111/anti.12522.

Machin, Amanda. "Democracy, Agony, and Rupture: A Critique of Climate Citizens' Assemblies." *Politische Vierteljahresschrift*, March 7, 2023. https://doi.org/10.1007/s11615-023-00455-5.

MacKinnon, Rebecca. "What to Get Right First: Five Things I Wish Web2 Had Addressed Early to Protect Human Rights." Starling Lab, September 2021. https://www.starlinglab.org/what-to-get-right-first/.

Madianou, Mirca. "Technocolonialism: Digital Innovation and Data Practices in the Humanitarian Response to Refugee Crises." *Social Media + Society* 5, no. 3 (April 1, 2019): 2056305119863146. https://doi.org/10.1177/2056305119863146.

Malloy, Judy. "The Origins of Social Media." In *Social Media Archeology and Poetics*, edited by Judy Malloy. Cambridge, MA: MIT Press, 2016.

Mannan, Morshed, and Nathan Schneider. "Exit to Community: Strategies for Multi-Stakeholder Ownership in the Platform Economy." *Georgetown Law Technology Review* 5, no. 1 (May 2021). https://georgetownlawtechreview.org/exit-to-community-strategies-for-multi-stakeholder-ownership-in-the-platform-economy/GLTR-05-2021/.

Manski, Sarah, and Ben Manski. "No Gods, No Masters, No Coders? The Future of Sovereignty in a Blockchain World." *Law and Critique* 29, no. 2 (July 1, 2018): 151–62. https://doi.org/10.1007/s10978-018-9225-z.

Marantz, Andrew. "Silicon Valley's Crisis of Conscience." *The New Yorker*, August 16, 2019. https://www.newyorker.com/magazine/2019/08/26/silicon-valleys-crisis-of-conscience.

Margonelli, Lisa. "Inside AOL's 'Cyber-Sweatshop.'" *Wired*, October 1, 1999. https://www.wired.com/1999/10/volunteers/.

Martins Rodrigues, Júlia, and Nathan Schneider. "Scaling Co-operatives through a Multi-Stakeholder Network: A Case Study in the Colorado Solar Energy Industry." *The Journal of Entrepreneurial and Organizational Diversity* 10, no. 2 (January 31, 2022): 29–53. https://doi.org/10.5947/jeod.2021.008.

Marx, Karl. *Capital*. Vol. 3. 1894. Reprint, Moscow: Progress Publishers, 1959.

———. *Grundrisse: Foundations of the Critique of Political Economy*. Harmondsworth, UK: Penguin, 1939. http://archive.org/details/in.ernet.dli.2015.505759.

Massanari, Adrienne. "#Gamergate and The Fappening: How Reddit's Algorithm, Governance, and Culture Support Toxic Technocultures." *New Media & Society* 19, no. 3 (March 1, 2017): 329–46. https://doi.org/10.1177/1461444815608807.

Mathur, Arunesh, Mihir Kshirsagar, and Jonathan Mayer. "What Makes a Dark Pattern . . . Dark? Design Attributes, Normative Considerations, and Measurement Methods." In *CHI '21: Proceedings of the 2021 CHI Conference on Human Factors in Computing Systems*, 1–18. New York: Association for Computing Machinery, 2021. https://doi.org/10.1145/3411764.3445610.

Matias, J. Nathan. "Going Dark: Social Factors in Collective Action against Platform Operators in the Reddit Blackout." In *CHI '16: Proceedings of the 2016 CHI Conference on Human Factors in Computing Systems*, 1138–51. Santa Clara, CA: ACM Press, 2016. https://doi.org/10.1145/2858036.2858391.

———. "Quitting Facebook & Google: Why Exit Option Democracy Is the Worst Kind of Democracy." Personal blog, December 11, 2018. https://natematias.medium.com/https -medium-com-natematias-quitting-facebook-google-aaf8f4c8ofbf.

Matney, Lucas. "Twitter's Decentralized Future." *TechCrunch*, January 15, 2021. https:// techcrunch.com/2021/01/15/twitters-vision-of-decentralization-could-also-be-the -far-rights-internet-endgame/.

Matsusaka, John J. *Let the People Rule: How Direct Democracy Can Meet the Populist Challenge*. Princeton, NJ: Princeton University Press, 2022. https://press.princeton.edu/books /paperback/9780691199740/let-the-people-rule.

Maurer, Bill, Taylor C. Nelms, and Lana Swartz. "'When Perhaps the Real Problem Is Money Itself!': The Practical Materiality of Bitcoin." *Social Semiotics* 23, no. 2 (April 1, 2013): 261–77. https://doi.org/10.1080/10350330.2013.777594.

Maxigas, and Guillaume Latzko-Toth. "Trusted Commons: Why 'Old' Social Media Matter." *Internet Policy Review* 9, no. 4 (October 21, 2020). https://policyreview.info/articles /analysis/trusted-commons-why-old-social-media-matter.

May, Ann Mari, and Gale Summerfield. "Creating a Space Where Gender Matters: Elinor Ostrom (1933–2012) Talks with Ann Mari May and Gale Summerfield." *Feminist Economics* 18, no. 4 (October 1, 2012): 25–37. https://doi.org/10.1080/13545701.2012.739725.

Mayo, Ed. *A Short History of Co-operation and Mutuality*. Co-operatives UK, 2017. https://www.uk.coop/sites/default/files/uploads/attachments/a-short-history-of -cooperation-and-mutuality_ed-mayo-web_english.pdf.

McEwan, Gregor, and Carl Gutwin. "A Case Study of How a Reduction in Explicit Leadership Changed an Online Game Community." *Computer Supported Cooperative Work (CSCW)* 26, no. 4 (December 1, 2017): 873–925. https://doi.org/10.1007/s10606-017-9282-0.

McGillicuddy, Aiden, Jean-Gregoire Bernard, and Jocelyn Cranefield. "Controlling Bad Behavior in Online Communities: An Examination of Moderation Work." *ICIS 2016 Proceedings*, December 11, 2016. https://aisel.aisnet.org/icis2016/SocialMedia/Presentations/23.

McPherson, Tara. "US Operating Systems at Mid-Century: The Intertwining of Race and UNIX." In *Race after the Internet*, edited by Lisa Nakamura and Peter Chow-White, 27–43. London: Routledge, 2011.

Meadows, Donella. "Leverage Points: Places to Intervene in a System." The Sustainability Institute, 1999. https://donellameadows.org/wp-content/userfiles/Leverage_Points.pdf.

Medvedev, Alexey N., Renaud Lambiotte, and Jean-Charles Delvenne. "The Anatomy of Reddit: An Overview of Academic Research." In *Dynamics On and Of Complex Networks III*, edited by Fakhteh Ghanbarnejad, Rishiraj Saha Roy, Fariba Karimi, Jean-Charles Delvenne, and Bivas Mitra, 183–204. Springer Proceedings in Complexity. Cham, Switzerland: Springer International Publishing, 2019.

Meehan, Mary Beth, and Fred Turner. *Seeing Silicon Valley: Life inside a Fraying America*. Chicago: University of Chicago Press, 2021.

Megarry, Jessica. *The Limitations of Social Media Feminism: No Space of Our Own*. Cham, Switzerland: Palgrave Macmillan, 2020.

Megill, Colin. "Pol.is in Taiwan." *pol.is* (blog), July 3, 2016. https://blog.pol.is/pol-is-in-taiwan -da7570d372b5.

Mejias, Ulises Ali. "To Fight Data Colonialism, We Need a Non-Aligned Tech Movement." *Al Jazeera*, September 8, 2020. https://www.aljazeera.com/opinions/2020/9/8/to-fight-data -colonialism-we-need-a-non-aligned-tech-movement.

Merk, Frederick, and Lois Bannister Merk. *Manifest Destiny and Mission in American History: A Reinterpretation*. 1963. Reprint, Cambridge, MA: Harvard University Press, 1995.

Merwe, Rachel Lara van der. "Imperial Play." *Communication, Culture and Critique*, June 17, 2020. https://doi.org/10.1093/ccc/tcaa012.

Milan, Stefania, and Emiliano Treré. "Big Data from the South(s): Beyond Data Universalism." *Television & New Media*, April 11, 2019. https://doi.org/10.1177/1527476419837739.

Miller, Robert J. "American Indians, the Doctrine of Discovery, and Manifest Destiny." *Wyoming Law Review* 11, no. 2 (2011): 329–49.

Mishra, Pankaj. "Gandhi for the Post-Truth Age." *The New Yorker*, October 15, 2018. https://www.newyorker.com/magazine/2018/10/22/gandhi-for-the-post-truth-age.

MolochDAO. The Original Grant Giving DAO. https://molochdao.com/.

Murimi, Renita. "Governance in DAOs: Lessons in Composability from Primate Societies and Modular Software." *MIT Computational Law Report*, December 5, 2022. https://law.mit.edu/pub/governanceindaos/release/1.

Murray, Padmini Ray. "A 'Feminist' Server to Help People Own Their Own Data." *The Bastion*, August 12, 2022. https://thebastion.co.in/politics-and/tech/a-feminist-server-to-help-people-own-their-own-data/.

Musgrave, Tyler, Alia Cummings, and Sarita Schoenebeck. "Experiences of Harm, Healing, and Joy among Black Women and Femmes on Social Media." In *CHI '22: Proceedings of the 2022 CHI Conference on Human Factors in Computing Systems*, 1–17. New York: Association for Computing Machinery, 2022. https://doi.org/10.1145/3491102.3517608.

Myers West, Sarah. "Censored, Suspended, Shadowbanned: User Interpretations of Content Moderation on Social Media Platforms." *New Media & Society* 20, no. 11 (November 1, 2018): 4366–83. https://doi.org/10.1177/1461444818773059.

Nabben, Kelsie. "Web3 as 'Self-Infrastructuring': The Challenge Is How." *Big Data & Society* 10, no. 1 (January 1, 2023): 20539517231159002. https://doi.org/10.1177/20539517231159002.

Nabben, Kelsie, Novita Puspasari, Megan Kelleher, and Sadhana Sanjay. "Grounding Decentralised Technologies in Cooperative Principles: What Can 'Decentralised Autonomous Organisations' (DAOs) and Platform Cooperatives Learn from Each Other?" Working paper, December 6, 2021. SSRN, https://doi.org/10.2139/ssrn.3979223.

Nagy, Peter, and Gina Neff. "Imagined Affordance: Reconstructing a Keyword for Communication Theory." *Social Media + Society* 1, no. 2 (September 22, 2015). https://doi.org/10.1177/2056305115603385.

Nakamura, Lisa. "The Unwanted Labour of Social Media: Women of Colour Call Out Culture As Venture Community Management." *New Formations* 86, no. 86 (December 15, 2015): 106–12. https://doi.org/10.3898/NEWF.86.06.2015.

Nash, Nathan. "International Facebook 'Friends': Toward McLuhan's Global Village." *McMaster Journal of Communication* 5, no. 1 (2009). https://doi.org/10.15173/mjc.v5i0.241.

Newport, Cal. "TikTok and the Fall of the Social-Media Giants." *The New Yorker*, July 28, 2022. https://www.newyorker.com/culture/cultural-comment/tiktok-and-the-fall-of-the-social-media-giants.

Newton, Casey. "Facebook's Big New Experiment in Governance." *Platformer*, September 20, 2022. https://www.platformer.news/p/facebooks-big-new-experiment-in-governance.

Nextcloud. "German Federal Administration Relies on Nextcloud as a Secure File Exchange Solution." *Nextcloud* (blog), April 18, 2018. https://nextcloud.com/blog/german-federal-administration-relies-on-nextcloud-as-a-secure-file-exchange-solution.

Ng, Jason. *Blocked on Weibo: What Gets Suppressed on China's Version of Twitter (And Why)*. New York: New Press, 2013.

Nicholas, Tom. *VC: An American History*. Cambridge, MA: Harvard University Press, 2019.

Niemeyer, Simon. "Scaling Up Deliberation to Mass Publics: Harnessing Mini-Publics in a Deliberative System." In *Deliberative Mini-Publics: Involving Citizens in the Democratic Process*, edited by André Bächtiger, 177–202. Colchester, UK: ECPR Press, 2014.

Noble, Safiya Umoja. *Algorithms of Oppression: How Search Engines Reinforce Racism*. New York: NYU Press, 2018.

Nordin, Astrid Hanna Maria. "Decolonising Friendship," *AMITY: The Journal of Friendship Studies* 6, no. 1 (2020). https://doi.org/10.5518/AMITY/32.

Norton, Anne. *Wild Democracy: Anarchy, Courage, and Ruling the Law*. Heretical Thought. Oxford, New York: Oxford University Press, 2023.

Nover, Scott. "Jack Dorsey Texted Elon Musk to Say Twitter Never Should Have Been a Company." *Quartz*, September 30, 2022. https://qz.com/jack-dorsey-said-making-twitter-a-company-was-its-orig-1849603325.

Nozick, Robert. *Anarchy, State, and Utopia*. New York: Basic Books, 1974.

Nucera, Diana J, and Valeria Mogilevich. "Teaching Community Technology Handbook." Detroit, MI: Detroit Community Technology Project, 2017. https://detroitcommunitytech.org/system/tdf/librarypdfs/TeachingCommunityTech.pdf?file=1&type=node&id=53&force=.

Nunes, Rodrigo. *Neither Vertical nor Horizontal: A Theory of Political Organization*. New York: Verso Books, 2021.

Nurik, Chloe. "'Men Are Scum': Self-Regulation, Hate Speech, and Gender-Based Censorship on Facebook." *International Journal of Communication* 13 (June 30, 2019). https://ijoc.org/index.php/ijoc/article/view/9608.

Nuth, Joan M. "Two Medieval Soteriologies: Anselm of Canterbury and Julian of Norwich." *Theological Studies* 53, no. 4 (December 1992): 611–45. https://doi.org/10.1177/004056399205300402.

O'Mara, Margaret. *The Code: Silicon Valley and the Remaking of America*. Ill. ed. New York: Penguin Press, 2019.

Oakland, Abby. "Minnesota's Digital Divide: How Minnesota Can Replicate the Rural Electrification Act to Deliver Rural Broadband." *Minnesota Law Review*, no. 105 (2020): 429. https://minnesotalawreview.org/article/minnesotas-digital-divide-how-minnesota-can-replicate-the-rural-electrification-act-to-deliver-rural-broadband/.

Oever, Niels ten. "The Metagovernance of Internet Governance." In *Power and Authority in Internet Governance*, edited by Blayne Haggart, Natasha Tusikov, and Jan Aart Scholte. London: Routledge, 2021. https://ebrary.net/174584/computer_science/metagovernance_internet_governance.

OneNet. "The OneNet Member Constitution." Wayback Machine archive, February 2, 1999. https://web.archive.org/web/19980623215303fw_/http://www.onenet.org/Cons/Intro.htm.

Ostrom, Elinor. "Beyond Markets and States: Polycentric Governance of Complex Economic Systems." *The American Economic Review* 100, no. 3 (2010): 641–72. https://www.jstor.org/stable/27871226.

———. *Governing the Commons: The Evolution of Institutions for Collective Action*. Cambridge, UK: Cambridge University Press, 1990.

———. *Understanding Institutional Diversity*. Princeton, NJ: Princeton University Press, 2006.

Ostrom, Vincent. "Artisanship and Artifact." *Public Administration Review* 40, no. 4 (1980): 309–17. https://doi.org/10.2307/3110256.

Ovadya, Aviv. "Towards Platform Democracy: Policymaking Beyond Corporate CEOs and Partisan Pressure." Cambridge, MA: Belfer Center for Science and International Affairs, Harvard University, October 18, 2021. https://www.belfercenter.org/publication /towards-platform-democracy-policymaking-beyond-corporate-ceos-and-partisan -pressure.

Owocki, Kevin. "A Brief History of Gitcoin from 2017–2022." Gitcoin Governance, December 15, 2021. https://gov.gitcoin.co/t/a-brief-history-of-gitcoin-from-2017-2022/9431.

Pan, Christina A., Sahil Yakhmi, Tara P. Iyer, Evan Strasnick, Amy X. Zhang, and Michael S. Bernstein. "Comparing the Perceived Legitimacy of Content Moderation Processes: Contractors, Algorithms, Expert Panels, and Digital Juries." *Proceedings of the ACM on Human-Computer Interaction* 6, no. CSCW1 (2022). https://doi.org/10.1145/3512929.

Pandey, Dave, and Erica Lawler. "Elon Musk on What the First Mars Colony Will Look Like." *Axios*, March 11, 2018. https://www.axios.com/2018/03/11/elon-musk-warns-about -ai-describes-plans-for-mars-colony-at-sxsw-1520795066.

Papacharissi, Zizi. *Affective Publics: Sentiment, Technology, and Politics*. New York: Oxford University Press, 2015.

———. *After Democracy: Imagining Our Political Future*. New Haven, CT: Yale University Press, 2021.

Parikka, Jussi. *What Is Media Archaeology?* London: Polity Press, 2012.

Park, Sungmee, and Sundaresan Jayaraman. "Textiles and Computing: Background and Opportunities for Convergence." In *Proceedings of the 2001 International Conference on Compilers, Architecture, and Synthesis for Embedded Systems*, 186–87. CASES '01. New York, NY, USA: Association for Computing Machinery, 2001. https://doi.org /10.1145/502217.502249.

Parks, Lisa. "Around the Antenna Tree: The Politics of Infrastructural Visibility." In *ACM SIGGRAPH 2007 Art Gallery*, 345. SIGGRAPH '07. New York: Association for Computing Machinery, 2007. https://doi.org/10.1145/1280120.1280266.

———. ""Stuff You Can Kick": Toward a Theory of Media Infrastructures." In *Between Humanities and the Digital*, edited by Patrik Svensson and David Theo Goldberg. Cambridge, MA: MIT Press, 2015. http://raley.english.ucsb.edu/wp-content/Engl800 /Parks-infrastructures.pdf.

Parler. "What Is the Parler Community Jury?" *Parler* (blog), December 1, 2022. https://web .archive.org/web/20230407204253/https://blog.parler.com/what-is-the-parler-commu nity-jury/.

Parvin, Phil. "Democracy without Participation: A New Politics for a Disengaged Era." *Res Publica* 24, no. 1 (February 2018): 31–52. https://doi.org/10.1007/s11158-017-9382-1.

Pateman, Carole. *Participation and Democratic Theory*. Cambridge, UK: Cambridge University Press, 1970.

Paul, Sanjukta. "Antitrust as Allocator of Coordination Rights." *UCLA Law Review* 67, no. 2 (2020). https://www.uclalawreview.org/antitrust-as-allocator-of-coordination-rights/.

Payne, Samuel B. "The Iroquois League, the Articles of Confederation, and the Constitution." *The William and Mary Quarterly* 53, no. 3 (1996): 605–20. https://doi.org/10.2307/2947207.

Pentzien, Jonas. "Political and Legislative Drivers and Obstacles for Platform Cooperativism in the United States, Germany, and France." Institute for the Cooperative Digital Economy, 2020.

Persily, Nathaniel, and Joshua A. Tucker, eds. *Social Media and Democracy: The State of the Field, Prospects for Reform*. Cambridge, UK: Cambridge University Press, 2020.

Pfaffenberger, Bryan. "'If I Want It, It's OK': Usenet and the (Outer) Limits of Free Speech." *The Information Society* 12, no. 4 (November 1996): 365–86. https://doi.org/10.1080 /019722496129350.

Phillips, Whitney. *This Is Why We Can't Have Nice Things: Mapping the Relationship between Online Trolling and Mainstream Culture*. Cambridge, MA: MIT Press, 2015.

Phillips, Whitney, and Ryan M. Milner. *The Ambivalent Internet: Mischief, Oddity, and Antagonism Online*. Cambridge, UK: Polity, 2017.

Pickard, Victor W. "United yet Autonomous: Indymedia and the Struggle to Sustain a Radical Democratic Network." *Media, Culture & Society* 28, no. 3 (May 1, 2006): 315–36. https://doi.org/10.1177/0163443706061685.

Pinto, Renata Ávila. "Digital Sovereignty or Digital Colonialism?" *Sur* 15, no. 27 (2018): 13.

Plantin, Jean-Christophe, Carl Lagoze, Paul N Edwards, and Christian Sandvig. "Infrastructure Studies Meet Platform Studies in the Age of Google and Facebook." *New Media & Society* 20, no. 1 (January 1, 2018): 293–310. https://doi.org/10.1177/1461444816661553.

Plantin, Jean-Christophe, and Gabriele de Seta. "WeChat as Infrastructure: The Techno-Nationalist Shaping of Chinese Digital Platforms." *Chinese Journal of Communication* 12, no. 3 (July 3, 2019): 257–73. https://doi.org/10.1080/17544750.2019.1572633.

Pogrebinschi, Thamy. *Innovating Democracy?: The Means and Ends of Citizen Participation in Latin America*. Cambridge Elements. Cambridge, UK: Cambridge University Press, 2023.

Pogue, James. "Inside the New Right's Next Frontier: The American West." *Vanity Fair*, February 21, 2023. https://www.vanityfair.com/news/2023/02/new-right-civil-war.

Power, Marilyn. "Social Provisioning as a Starting Point for Feminist Economics." *Feminist Economics* 10, no. 3 (November 1, 2004): 3–19. https://doi.org/10.1080/13545700420002 67608.

Procaccia, Ariel. "Citizens' Assemblies Are Upgrading Democracy: Fair Algorithms Are Part of the Program." *Scientific American*, November 1, 2022. https://doi.org/10.1038 /scientificamerican1122-52.

Puar, Jasbir K. "'I Would Rather Be a Cyborg Than a Goddess': Becoming Intersectional in Assemblage Theory." *philoSOPHIA* 2, no. 1 (2012): 49–66. https://muse.jhu.edu/article /486621/pdf.

Putnam, Robert D., and Shaylyn Romney Garrett. *The Upswing: How America Came Together a Century Ago and How We Can Do It Again*. New York: Simon & Schuster, 2020.

Putnam, Robert D., Robert Leonardi, and Raffaella Y. Nanetti. *Making Democracy Work: Civic Traditions in Modern Italy*. Princeton, NJ: Princeton University Press, 1992.

Qiu, Jack Linchuan. *Goodbye iSlave: A Manifesto for Digital Abolition*. University of Illinois Press, 2017.

Quijano, Aníbal. "Coloniality and Modernity/Rationality." *Cultural Studies* 21, no. 2–3 (March 2007): 168–78. https://doi.org/10.1080/09502380601164353.

Radebaugh, Jacqueline, and Yev Muchnik. "Solving the Riddle of the DAO with Colorado's Co-operative Laws." *The Defiant*, December 16, 2021. https://thedefiant.io/solving-the-riddle -of-the-dao-with-colorados-cooperative-laws/.

Ramirez, Dennis, Jenny Saucerman, and Jeremy Dietmeier. "Twitch Plays Pokemon: A Case Study in Big G Games." In *Proceedings of DiGRA*, 2014.

Rankin, Joy Lisi. *A People's History of Computing in the United States*. Cambridge, MA: Harvard University Press, 2018.

Rao, Parimala V. "Gandhi, Untouchability and the Postcolonial Predicament: A Note." *Social Scientist* 37, no. 1/2 (2009): 64–70. https://www.jstor.org/stable/27644310.

Reagle, Joseph Michael. *Good Faith Collaboration: The Culture of Wikipedia*. Cambridge, MA: MIT Press, 2010. https://reagle.org/joseph/2010/gfc/.

Reid, Elizabeth M. "Communication and Community on Internet Relay Chat: Constructing Communities." In *High Noon on the Electronic Frontier: Conceptual Issues in Cyberspace*, edited by Peter Ludlow. Cambridge, MA: MIT Press, 1996.

Reijers, Wessel, Iris Wuisman, Morshed Mannan, Primavera De Filippi, Christopher Wray, Vienna Rae-Looi, Angela Cubillos Vélez, and Liav Orgad. "Now the Code Runs Itself: On-Chain and Off-Chain Governance of Blockchain Technologies." *Topoi*, December 17, 2018. https://doi.org/10.1007/s11245-018-9626-5.

Resnick, Pete. "On Consensus and Humming in the IETF." Request for Comments. Internet Engineering Task Force, June 2014. https://doi.org/10.17487/RFC7282.

Restakis, John. *Civilizing the State: Reclaiming Politics for the Common Good*. Gabriola Island, BC: New Society Publishers, 2021.

Reybrouck, David Van. *Against Elections*. New York: Seven Stories Press, 2018.

Rheingold, Howard. "Online Community Governance: Six Case Studies." Patreon, February 13, 2021. https://www.patreon.com/posts/47388851.

———. *The Virtual Community: Homesteading on the Electronic Frontier*. Reading, MA: Addison-Wesley, 1993. https://people.well.com/user/hlr/vcbook/.

———. *The Virtual Community: Homesteading on the Electronic Frontier*. Rev. ed. Cambridge, MA: MIT Press, 2000.

Rintel, E. Sean, and Jeffery Pittam. "Strangers in a Strange Land: Interaction Management on Internet Relay Chat." *Human Communication Research* 23, no. 4 (June 1997): 507–34. https://doi.org/10.1111/j.1468-2958.1997.tb00408.x.

Robbins, Robert B., Cindy V. Schlaefer, and Jessica Lutrin. "From Home Sharing and Ride Sharing to Shareholding." Pillsbury Law, October 25, 2018. https://www.pillsburylaw.com/en/news-and-insights/rule-701-revision-uber-airbnb.html.

Rochefort, Alex. "Regulating Social Media Platforms: A Comparative Policy Analysis." *Communication Law and Policy* 25, no. 2 (April 2, 2020): 225–60. https://doi.org/10.1080/10811680.2020.1735194.

Ronzhyn, Alexander, Ana Sofia Cardenal, and Albert Batlle Rubio. "Defining Affordances in Social Media Research: A Literature Review." *New Media & Society*, November 23, 2022. https://doi.org/10.1177/14614448221135187.

Roscam Abbing, Roel. "On Cultivating the Installable Base." In *Proceedings of the Participatory Design Conference 2022—Volume 2*, 203–7. PDC '22. New York, NY, USA: Association for Computing Machinery, 2022. https://doi.org/10.1145/3537797.3537875.

Roy, Arundhati. *The Doctor and the Saint: Caste, Race, and Annihilation of Caste: The Debate Between B. R. Ambedkar and M. K. Gandhi*. Chicago: Haymarket Books, 2017.

Rubenstein, Mary-Jane, and Jenna Supp-Montgomerie. "Somewhere Out There: Corporate Utopias of Space and Sea." *The Immanent Frame*, April 30, 2021. https://tif.ssrc.org/2021/04/30/somewhere-out-there/.

Russ, Joanna. "SF and Technology as Mystification." *Science Fiction Studies* 5, no. 3 (November 1978). https://www.jstor.org/stable/4239201.

Russell, Bertie. "Beyond the Local Trap: New Municipalism and the Rise of the Fearless Cities." *Antipode* 51, no. 3 (2019): 989–1010. https://doi.org/10.1111/anti.12520.

Russo, Camila. *The Infinite Machine: How an Army of Crypto-hackers Is Building the Next Internet with Ethereum.* New York, NY: Harper Business, 2020.

Sabetti, Filippo. "Constitutional Artisanship and Institutional Diversity: Elinor Ostrom, Vincent Ostrom, and the Workshop." *The Good Society* 20, no. 1 (2011): 73–83. https://doi.org/10.5325/goodsociety.20.1.0073.

Satz, Debra. *Why Some Things Should Not Be for Sale: The Moral Limits of Markets.* New York: Oxford University Press, 2010.

Savic, Selena, and Stefanie Wuschitz. "Feminist Hackerspace as a Place of Infrastructure Production." *Ada: A Journal of Gender, New Media, and Technology,* no. 13 (2018). https://adanewmedia.org/2018/05/issue13-savic-wuschitz/.

Savolainen, Laura. "The Shadow Banning Controversy: Perceived Governance and Algorithmic Folklore." *Media, Culture & Society* 44, no. 6 (September 1, 2022): 1091–1109. https://doi.org/10.1177/01634437221077174.

Scarry, Elaine. *Thermonuclear Monarchy: Choosing between Democracy and Doom.* Summary ed. New York: W. W. Norton, 2016.

Schneider, Nathan. "Cryptoeconomics as a Limitation on Governance." Mirror, August 11, 2022. https://ntnsndr.mirror.xyz/zO27EOn9P_62jVlautpZD5hHB7ycf3Cfc2N6byz6DOk.

———. "Decentralization: An Incomplete Ambition." *Journal of Cultural Economy* 12, no. 4 (April 17, 2019): 1–21. https://doi.org/10.1080/17530350.2019.1589553.

———. "Democrats and Republicans Both Have a Big Blind Spot on Net Neutrality." *Quartz,* December 21, 2017. https://qz.com/1163274/net-neutrality-debate-the-democrats-have-one-big-blind-spot/.

———. "Digital Kelsoism: Employee Stock Ownership as a Pattern for the Online Economy." In *Reimagining the Governance of Work and Employment,* edited by Dionne Pohler, 234–46. Ithaca, NY: Cornell University Press, 2020. https://osf.io/7wrab/.

———. "An Economy for Anything." *Plough,* October 23, 2019. https://www.plough.com/en/topics/justice/social-justice/economic-justice/an-economy-for-anything.

———. "Enabling Community-Owned Platforms: A Proposal for a Tech New Deal." In *Regulating Big Tech: Policy Responses to Digital Dominance,* edited by Martin Moore and Damian Tambini. Oxford University Press, 2021. https://ntnsndr.in/TechNewDeal.

———. *Everything for Everyone: The Radical Tradition That Is Shaping the Next Economy.* New York: Nation Books, 2018.

———. "The Future of Owning the Internet." *Vice,* March 24, 2016. https://www.vice.com/en/article/exq3ne/zen-and-the-art-of-internet-maintenance-v23n1.

———. "Governable Stacks against Digital Colonialism." *tripleC: Communication, Capitalism & Critique* 20, no. 1 (January 12, 2022): 19–36. https://doi.org/10.31269/triplec.v20i1.1281.

———. "How OWS' 'Anti-Market Research Analyst' Helps the Movement Go Viral." Waging Nonviolence, July 1, 2012. https://wagingnonviolence.org/2012/07/how-ows-anti-market-research-analyst-helps-the-movement-go-viral/.

———. "How We Can Encode Human Rights in the Blockchain." *Noema,* June 7, 2022. https://www.noemamag.com/how-we-can-encode-human-rights-in-the-blockchain.

———. "The Joy of Slow Computing." *The New Republic*, May 20, 2015. https://newrepublic
.com/article/121832/pleasure-do-it-yourself-slow-computing.

———. "Lighten the Load of the Nation-State." *Kernel Magazine*, 2022. https://www.kernelmag
.io/2/lighten-nation-state.

———. "Mediated Ownership: Capital as Media." *Media, Culture & Society* 42, no. 3 (April 1,
2020): 449–59. https://doi.org/10.1177/0163443719899035.

———. "Startups Need a New Option: Exit to Community." Hacker Noon, September 16, 2019.
https://hackernoon.com/startups-need-a-new-option-exit-to-community-ig12v2z73.

———. *Thank You, Anarchy: Notes from the Occupy Apocalypse by Nathan Schneider*. Berkeley: University of California Press, 2013.

———. "The Tyranny of Openness: What Happened to Peer Production?" *Feminist Media
Studies* 22, no. 6 (2022). https://doi.org/10.1080/14680777.2021.1890183.

———. "'Truly, Much Can Be Done': Cooperative Economics from the Book of Acts to Pope
Francis." In *Care for the World: Laudato Si' and Catholic Social Thought in an Era of
Climate Crisis*, edited by Frank Pasquale. Cambridge, UK: Cambridge University Press,
2019. https://osf.io/jhrmd.

Schneider, Nathan, Primavera De Filippi, Seth Frey, Joshua Z. Tan, and Amy X. Zhang.
"Modular Politics: Toward a Governance Layer for Online Communities." *Proceedings
of the ACM on Human-Computer Interaction*, April 2021. https://dl.acm.org/doi
/10.1145/3449090.

Schneider, Nathan, and Amy Hasinoff. "Mastodon Isn't Just a Replacement for Twitter." *Noema*,
November 29, 2022. https://www.noemamag.com/mastodon-isnt-just-a-replacement
-for-twitter.

Schoenebeck, Sarita, Oliver L Haimson, and Lisa Nakamura. "Drawing from Justice
Theories to Support Targets of Online Harassment." *New Media & Society* 23, no. 5 (May
1, 2021): 1278–1300. https://doi.org/10.1177/1461444820913122.

Scholz, Trebor, and Nathan Schneider. *Ours to Hack and to Own: The Rise of Platform
Cooperativism, a New Vision for the Future of Work and a Fairer Internet*. New York: OR
Books, 2016.

Schor, Juliet. *After the Gig: How the Sharing Economy Got Hijacked and How to Win It Back*.
Berkeley, CA: University of California Press, 2020.

Schwarzenbach, Sibyl A. "Democracy and Friendship." *Journal of Social Philosophy* 36, no. 2
(May 2005): 233–54. https://doi.org/10.1111/j.1467-9833.2005.00269.x.

Seering, Joseph. "Reconsidering Self-Moderation: The Role of Research in Supporting
Community-Based Models for Online Content Moderation." *Proceedings of the ACM
on Human-Computer Interaction* 4 (October 14, 2020): 1–28. https://doi.org/10.1145
/3415178.

Seering, Joseph, Tony Wang, Jina Yoon, and Geoff Kaufman. "Moderator Engagement and
Community Development in the Age of Algorithms." *New Media & Society*, January 11,
2019. https://doi.org/10.1177/1461444818821316.

Sengupta, Rakesh. "Towards a Decolonial Media Archaeology: The Absent Archive of
Screenwriting History and the Obsolete Munshi." *Theory, Culture & Society* 38, no. 1
(January 1, 2021): 3–26. https://doi.org/10.1177/0263276420930276.

Serafini, Paula. "Community Radio as a Space of Care: An Ecofeminist Perspective on Media Production in Environmental Conflicts." *International Journal of Communication* 13
(December 1, 2019). https://ijoc.org/index.php/ijoc/article/view/11524.

Sharlet, Jeff. "'He's the Chosen One to Run America': Inside the Cult of Trump, His Rallies Are Church and He Is the Gospel." *Vanity Fair*, June 18, 2020. https://www.vanityfair.com /news/2020/06/inside-the-cult-of-trump-his-rallies-are-church-and-he-is-the-gospel.

Shaw, Aaron, and Benjamin M. Hill. "Laboratories of Oligarchy? How the Iron Law Extends to Peer Production." *Journal of Communication* 64, no. 2 (April 1, 2014): 215–38. https:// doi.org/10.1111/jcom.12082.

Sheldrake, Merlin. *Entangled Life: How Fungi Make Our Worlds, Change Our Minds and Shape Our Futures*. New York: Random House, 2020.

Sheldrake, Philip. "Michel de Certeau: Spirituality and the Practice of Everyday Life." *Spiritus: A Journal of Christian Spirituality* 12, no. 2 (2012): 207–16. https://doi.org/10.1353 /scs.2012.0024.

Sherman, Justin. "Don't Be Fooled by Big Tech's Anti-China Sideshow." *Wired*, July 30, 2020. https://www.wired.com/story/opinion-dont-be-fooled-by-big-techs-anti-china-sideshow/.

Siddarth, Divya, Danielle Allen, and E. Glen Weyl. "The Web3 Decentralization Debate Is Focused on the Wrong Question." *Wired*, May 12, 2022. https://www.wired.com/story /web3-blockchain-decentralization-governance/.

Siddarth, Divya, Sergey Ivliev, Santiago Siri, and Paula Berman. "Who Watches the Watchmen? A Review of Subjective Approaches for Sybil-Resistance in Proof of Personhood Protocols." *Frontiers in Blockchain* 3 (2020): 46. https://doi.org/10.3389/fbloc.2020.590171.

Siegel, Jacob. "The Red-Pill Prince." *Tablet Magazine*, March 31, 2022. https://www.tabletmag .com/sections/news/articles/red-pill-prince-curtis-yarvin.

Silva-Leander, Annika. *Global State of Democracy Report 2021: Building Resilience in a Pandemic Era*. Stockholm: International IDEA, November 22, 2021. https://www.idea.int /gsod-2021/.

Sitrin, Marina, Dario Azzellini, and David Harvey. *They Can't Represent Us!: Reinventing Democracy from Greece to Occupy*. London: Verso, 2014.

Smith, Andrea. "Abolition Feminism and Jumping Scale: Transformative Justice as a Way of Life." In *Feminism—Corporeality, Materialism, and Beyond*, edited by Dennis Erasga and Michael Eduard Labayandoy. London: IntechOpen, 2023. https://doi.org/10.5772 /intechopen.110553.

Smith, Linda Tuhiwai. *Decolonizing Methodologies: Research and Indigenous Peoples*. London: Zed Books, 1999.

Soleimani, Ameen, Arjun Bhuptani, James Young, Layne Haber, and Rahul Sethuram. "The Moloch DAO: Beating the Tragedy of the Commons Using Decentralized Autonomous Organizations," 2019. https://raw.githubusercontent.com/MolochVentures/Whitepaper /master/Whitepaper.pdf.

Sørensen, Eva, and Jacob Torfing. "Theoretical Approaches to Metagovernance." In *Theories of Democratic Network Governance*, edited by Eva Sørensen and Jacob Torfing, 169–82. London: Palgrave Macmillan UK, 2007. https://doi.org/10.1057/9780230625006_10.

Sousa Santos, Boaventura de. *Epistemologies of the South: Justice against Epistemicide*. 2014. Reprint, London: Routledge, 2016.

———. "Participatory Budgeting in Porto Alegre: Toward a Redistributive Democracy." *Politics & Society* 26, no. 4 (December 1998): 461–510. https://doi.org/10.1177/0032329 298026004003.

Spencer, Henry, and David Lawrence. *Managing Usenet*. Sebastopol, CA: O'Reilly & Associates, 1998.

Spicer, Jason. "Cooperative Enterprise at Scale: Comparative Capitalisms and the Political Economy of Ownership." *Socio-Economic Review*, October 15, 2021. https://doi.org/10.1093/ser/mwab010.

Srinivasan, Balaji. *The Network State*, 2022, online book. https://thenetworkstate.com/.

SSL Nagbot. "Feminist Hacking/Making: Exploring New Gender Horizons of Possibility." *Journal of Peer Production*, no. 8 (March 2016). http://peerproduction.net/issues/issue-8-feminism-and-unhacking-2/feminist-hackingmaking-exploring-new-gender-horizons-of-possibility/.

Stark, Josh. "Making Sense of Cryptoeconomics." *CoinDesk*, August 19, 2017. https://www.coindesk.com/making-sense-cryptoeconomics.

Stasavage, David. *The Decline and Rise of Democracy: A Global History from Antiquity to Today*. Princeton, NJ: Princeton University Press, 2020.

Stempeck, Matt. "Next-Generation Engagement Platforms, and How They Are Useful Right Now (Part 1)." Civicist, May 12, 2020. https://web.archive.org/web/20201101012224/https://civichall.org/civicist/next-generation-engagement-platforms-and-how-are-they-useful-right-now-part-1/.

Stewart, Emily. "Facebook Will Never Strip Away Mark Zuckerberg's Power." *Vox*, May 30, 2019. https://www.vox.com/recode/2019/5/30/18644755/facebook-stock-shareholder-meeting-mark-zuckerberg-vote.

Stout, Jeffrey. *Democracy and Tradition*. Princeton, NJ: Princeton University Press, 2004.

Strober, Myra H. "Rethinking Economics through a Feminist Lens." *The American Economic Review* 84, no. 2 (1994): 143–47.

Suler, John. "The Online Disinhibition Effect." *CyberPsychology & Behavior* 7, no. 3 (2004): 321–26. https://doi.org/10.1089/1094931041291295.

Sullivan, Tim. "Blitzscaling." *Harvard Business Review*, April 1, 2016. https://hbr.org/2016/04/blitzscaling.

Sunstein, Cass R. *#Republic: Divided Democracy in the Age of Social Media*. Princeton, NJ: Princeton University Press, 2018.

Supp-Montgomerie, Jenna. "Infrastructural Awareness." *Cultural Studies*, October 20, 2021. https://doi.org/10.1080/09502386.2021.1988121.

Swann, Thomas. *Anarchist Cybernetics: Control and Communication in Radical Politics*. Bristol, UK: Bristol University Press, 2020.

Swartz, Lana. "Blockchain Dreams: Imagining Techno-Economic Alternatives after Bitcoin." In *Another Economy Is Possible: Culture and Economy in a Time of Crisis*, edited by Manuel Castells, 82–105. Cambridge, UK: Polity, 2017.

———. *New Money: How Payment Became Social Media*. New Haven, CT: Yale University Press, 2020.

———. "Theorizing the 2017 Blockchain ICO Bubble as a Network Scam." *New Media & Society* 24, no. 7 (July 1, 2022): 1695–1713. https://doi.org/10.1177/14614448221099224.

Tait, Joshua. "Mencius Moldbug and Neoreaction." In *Key Thinkers of the Radical Right: Behind the New Threat to Liberal Democracy*, edited by Mark Sedgwick. New York: Oxford University Press, 2019. https://doi.org/10.1093/oso/9780190877583.003.0012.

Talbot, David A., Kira Hope Hessekiel, and Danielle Leah Kehl. "Community-Owned Fiber Networks: Value Leaders in America." Berkman Klein Center for Internet & Society Research Publication, Harvard University, 2017. https://dash.harvard.edu/handle/1/34623859.

Tan, Joshua, Max Langenkamp, Anna Weichselbraun, Ann Brody, and Lucia Korpas. "Constitutions of Web3." Metagovernance Project, 2022. https://constitutions.metagov.org/article.

Tang, Xiaobing. *Chinese Modern: The Heroic and the Quotidian.* Durham, NC: Duke University Press, 2000.

Tarnoff, Ben. *The Making of the Tech Worker Movement.* Logic Magazine, 2020. https://logicmag.io/the-making-of-the-tech-worker-movement/full-text/.

Taylor, Keith. "An Analysis of the Entrepreneurial Institutional Ecosystems Supporting the Development of Hybrid Organizations: The Development of Cooperatives in the U.S." *Journal of Environmental Management,* 2021, 8.

Temprano, Victor. "Native Land: Social Media Education and Community Voices." In *Digital Mapping and Indigenous America,* edited by Janet Berry Hess. London: Routledge, 2021.

Terranova, Tiziana. "Free Labor: Producing Culture for the Digital Economy." *Social Text* 18 (June 1, 2000): 33–58. https://doi.org/10.1215/01642472-18-2_63-33.

———. "Red Stack Attack! Algorithms, Capital and the Automation of the Common." *Effimera,* March 8, 2014. http://effimera.org/red-stack-attack-algorithms-capital-and-the-automation-of-the-common-di-tiziana-terranova/.

Thaning, Morten Sørensen, Marius Gudmand-Høyer, and Sverre Raffnsøe. "Ungovernable: Reassessing Foucault's Ethics in Light of Agamben's Pauline Conception of Use." *International Journal of Philosophy and Theology* 77, no. 3 (May 26, 2016): 191–218. https://doi.org/10.1080/21692327.2016.1235987.

Thatcher, Jim, David O'Sullivan, and Dillon Mahmoudi. "Data Colonialism through Accumulation by Dispossession: New Metaphors for Daily Data." *Environment and Planning D: Society and Space* 34, no. 6 (December 1, 2016): 990–1006. https://doi.org/10.1177/0263775816633195.

The WELL. "How The Well Bought Itself," 2012. https://www.well.com/about-2/pr/how-the-well-bought-itself/.

Thiel, Peter, and Blake Masters. *Zero to One: Notes on Startups, or How to Build the Future.* Ill. ed. New York: Currency, 2014.

Thompson, David J. "Frederick Douglass and Co-ops in 1846." *National Cooperative Bank* (blog), February 3, 2022. https://www.ncb.coop/blog/ncb-frederick-douglass-and-co-ops-in-1846.

Thompson, Margaret E., Katerina Anfossi Gómez, and María Suárez Toro. "Women's Alternative Internet Radio and Feminist Interactive Communications." *Feminist Media Studies* 5, no. 2 (July 1, 2005): 215–36. https://doi.org/10.1080/14680770500124306.

Tocqueville, Alexis de. *Democracy in America.* Translated by Henry Reeve. 2 vols. London: Saunders and Otley, 1840. https://www.gutenberg.org/ebooks/815 and https://www.gutenberg.org/ebooks/816.

Tomba, Luigi. "Of Quality, Harmony, and Community: Civilization and the Middle Class in Urban China." *Positions: Asia Critique* 17, no. 3 (August 1, 2009): 591–616. https://doi.org/10.1215/10679847-2009-016.

Torfing, Jacob. "Metagovernance." In *Handbook on Theories of Governance,* edited by Christopher Ansell and Jacob Torfing, 567–79. Cheltenham, UK: Edward Elgar, 2022. https://www.elgaronline.com/view/edcoll/9781800371965/9781800371965.00059.xml.

Törnberg, Petter. "How Digital Media Drive Affective Polarization through Partisan Sorting." *Proceedings of the National Academy of Sciences of the United States of America* 119, no. 42 (October 18, 2022). https://doi.org/10.1073/pnas.2207159119.

Toupin, Sophie. "Feminist Hackerspaces: The Synthesis of Feminist and Hacker Cultures." *Journal of Peer Production*, no. 5 (October 2014). http://peerproduction.net/issues/issue -5-shared-machine-shops/peer-reviewed-articles/feminist-hackerspaces-the-synthesis -of-feminist-and-hacker-cultures/.

Tourani, Parastou, Bram Adams, and Alexander Serebrenik. "Code of Conduct in Open Source Projects." In *2017 IEEE 24th International Conference on Software Analysis, Evolution and Reengineering (SANER)*, 24–33, 2017. https://doi.org/10.1109/SANER.2017 .7884606.

Tria Kerkvliet, Benedict J. "Everyday Politics in Peasant Societies (and Ours)." *Journal of Peasant Studies* 36, no. 1 (January 1, 2009): 227–43. https://doi.org/10.1080/03066150902 820487.

Troncoso, Stacco, and Ann Marie Utratel. *Groove Is in the Heart: The DiSCO Elements*. DisCO. coop, 2020. https://disco.coop/wp-content/uploads/2020/12/DisCO_Elements_v1-3 .pdf.

Tsai, Lily L. *Accountability without Democracy: Solidary Groups and Public Goods Provision in Rural China*. New York: Cambridge University Press, 2007.

Tseng, Yu-Shan. "Algorithmic Empowerment: A Comparative Ethnography of Two Open-Source Algorithmic Platforms Decide Madrid and vTaiwan." *Big Data & Society* 9, no. 2 (July 1, 2022). https://doi.org/10.1177/20539517221123505.

Tsing, Anna Lowenhaupt. *The Mushroom at the End of the World*. Princeton, NJ: Princeton University Press, 2015.

———. "On Nonscalability: The Living World Is Not Amenable to Precision-Nested Scales." *Common Knowledge* 18, no. 3 (August 1, 2012): 505–24. https://doi.org/10 .1215/0961754X-1630424.

Tuck, Eve, and K Wayne Yang. "Decolonization Is Not a Metaphor." *Decolonization: Indigeneity, Education & Society* 1, no. 1 (2012): 40.

Tufekci, Zeynep. *Twitter and Tear Gas: The Power and Fragility of Networked Protest*. New Haven, CT: Yale University Press, 2017.

Turnbull, Cadwell. "Monsters Come Howling in Their Season." *The Verge*, January 23, 2019, n.p. https://www.theverge.com/2019/1/23/18175285/cadwell-turnbull-sci-fi-story-common -ai-climate-change-better-worlds.

Turner, Fred. "The Arts at Facebook: An Aesthetic Infrastructure for Surveillance Capitalism." *Poetics* 67 (April 1, 2018): 53–62. https://doi.org/10.1016/j.poetic.2018.03.003.

———. "Burning Man at Google: A Cultural Infrastructure for New Media Production." *New Media & Society* 11, no. 1–2 (February 1, 2009): 73–94. https://doi.org/10.1177 /1461444808099575.

———. *From Counterculture to Cyberculture: Stewart Brand, the Whole Earth Network, and the Rise of Digital Utopianism*. Chicago: University of Chicago Press, 2010.

Vaheesan, Sandeep, and Nathan Schneider. "Cooperative Enterprise as an Antimonopoly Strategy." *Penn State Law Review* 124, no. 1 (2019): 1–55. http://www.pennstatelawreview .org/print-issues/cooperative-enterprise-as-an-antimonopoly-strategy/.

Vaidhyanathan, Siva. *Antisocial Media: How Facebook Disconnects Us and Undermines Democracy*. New York: Oxford University Press, 2018.

Vaughn, Paige E., Kyle Peyton, and Gregory A. Huber. "Mass Support for Proposals to Reshape Policing Depends on the Implications for Crime and Safety." *Criminology & Public Policy* 21, no. 1 (2022): 125–46. https://doi.org/10.1111/1745-9133.12572.

Vázquez, Ronaldo. "Towards a Decolonial Critique of Modernity: Buen Vivir, Relationality and the Task of Listening." In *Kapital, Armut, Entwicklung,* edited by Raúl Fornet-Betancourt. Denktraditionen im Dialog: Studien zur Befreiung und Interkulturalität 33. Aachen, Germany: Wissenschaftsverlag Mainz, 2012.

V-Dem Institute. *Democracy Report 2022: Autocratization Changing Nature?* Gothenburg, Sweden: University of Gothenburg, March 2022. https://v-dem.net/media/publications /dr_2022.pdf.

Vincent, Pheroze L. "Boards That Gave Artisans a Voice Scrapped." *The Telegraph India,* August 8, 2020. https://www.telegraphindia.com/india/all-india-handloom-board-and -the-all-india-handicrafts-board-scrapped/cid/1788527.

Vitalist International. "Life Finds a Way." *Commune,* January 1, 2019. https://communemag .com/life-finds-a-way/.

Wade, Keith. *The Anarchist's Guide to the BBS.* Port Townsend, WA: Loompanics Unlimited, 1990.

Wajcman, Judy. *Feminism Confronts Technology.* University Park: Penn State University Press, 1991.

Walden, Jesse. "Past, Present, Future: From Co-ops to Cryptonetworks." *Variant Fund* (blog), March 2, 2019. https://variant.fund/past-present-future-from-co-ops-to-crypto networks/.

Wang, Hannah M., Beril Bulat, Stephen Fujimoto, and Seth Frey. "Governing for Free: Rule Process Effects on Reddit Moderator Motivations," June 11, 2022. https://doi.org /10.48550/arXiv.2206.05629.

Wang, Xuan, Kasper Juffermans, and Caixia Du. "Harmony as Language Policy in China: An Internet Perspective." *Language Policy* 15, no. 3 (August 1, 2016): 299–321. https://doi .org/10.1007/s10993-015-9374-y.

Waring, Marilyn. *If Women Counted: A New Feminist Economics.* San Francisco: Harper & Row, 1988.

Watkin, Amy-Louise, Vivian Gerrand, and Maura Conway. "Introduction: Exploring Societal Resilience to Online Polarization and Extremism." *First Monday,* May 2, 2022. https://doi.org/10.5210/fm.v27i5.12595.

Webber, Stephen. "Introducing OpenZeppelin Governor." *OpenZeppelin* (blog), August 17, 2021. https://blog.openzeppelin.com/governor-smart-contract/.

Welch, Cheryl B. "Colonial Violence and the Rhetoric of Evasion: Tocqueville on Algeria." *Political Theory* 31, no. 2 (April 1, 2003): 235–64. https://doi.org/10.1177/0090591702251011.

Wershler, Darren, Lori Emerson, and Jussi Parikka. *The Lab Book: Situated Practices in Media Studies.* Minneapolis: University of Minnesota Press, 2022. https://manifold.umn.edu /projects/the-lab-book.

Weyl, E. Glen, Puja Ohlhaver, and Vitalik Buterin. "Decentralized Society: Finding Web3's Soul." SSRN, May 10, 2022. SSRN, https://doi.org/10.2139/ssrn.4105763.

Whittington, Oli. *Democratic Innovation and Digital Participation.* London: Nesta, September 2022.

Wike, Richard, Laura Silver, and Alexandra Castillo. *Many People around the World Are Unhappy with How Democracy Is Working.* Washington, DC: Pew Research Center, April 29, 2019. https://www.pewglobal.org/2019/04/29/many-across-the-globe-are-dissatisfied-with -how-democracy-is-working/.

Wikipedia. "Wikipedia:Role of Jimmy Wales." In *Wikipedia*, July 1, 2020. https://en .wikipedia.org/wiki/Wikipedia:Role_of_Jimmy_Wales.

Will, George F. "Sympathy For Guinier." *Newsweek*, June 13, 1993. https://www.newsweek .com/sympathy-guinier-194016.

Winger-Bearskin, Amelia. "Before Everyone Was Talking about Decentralization, Decentralization Was Talking to Everyone." *Immerse*, July 2, 2018. https://immerse.news /decentralized-storytelling-d8450490b3ee.

Winner, Langdon. "Do Artifacts Have Politics?" *Daedalus* 109, no. 1 (Winter 1980). https:// www.jstor.org/stable/20024652.

Wittkower, D. E. "Principles of Anti-discriminatory Design." In *2016 IEEE International Symposium on Ethics in Engineering, Science and Technology (ETHICS)*, 1–7. https://doi.org /10.1109/ETHICS.2016.7560055.

Woessner, Matthew. "Teaching with SimCity: Using Computer Games to Construct Dynamic Governance Simulations." Paper presented at APSA Teaching and Learning Conference, Long Beach, CA, February 8–10, 2013. SSRN, posted February 9, 2020, https:// doi.org/10.2139/ssrn.2209488.

Wójtowicz, Tomasz, and Konrad Szocik. "Democracy or What? Political System on the Planet Mars after Its Colonization." *Technological Forecasting and Social Change* 166 (May 1, 2021). https://doi.org/10.1016/j.techfore.2021.120619.

Wolfson, Todd. *Digital Rebellion: The Birth of the Cyber Left*. Urbana, IL: University of Illinois Press, 2014.

Wooley, David R. "PLATO: The Emergence of Online Community." In *Social Media Archeology and Poetics*, edited by Judy Malloy. Cambridge, MA: MIT Press, 2016.

Wright, Del Jr. "Quadratic Voting and Blockchain Governance." *UMKC Law Review* 88, no. 2 (2019–2020): 475–96.

Wright, Erik Olin. *Envisioning Real Utopias*. London: Verso, 2010.

Wu, Sherry Jueyu, and Elizabeth Levy Paluck. "Participatory Practices at Work Change Attitudes and Behavior toward Societal Authority and Justice." *Nature Communications* 11, no. 2633 (May 26, 2020). https://doi.org/10.1038/s41467-020-16383-6.

Wyatt, Sally. "Technological Determinism Is Dead; Long Live Technological Determinism." In *The Handbook of Science and Technology Studies*, edited by Edward J. Hackett and Society for Social Studies of Science, 3rd ed. Cambridge, MA: MIT Press, 2008.

Yang, Bin. *Cowrie Shells and Cowrie Money: A Global History*. London: Routledge, 2018. https://doi.org/10.4324/9780429489587.

Youngs, Richard, and Ken Godfrey. "Democratic Innovations from around the World: Lessons for the West." Carnegie Europe, November 3, 2022. https://carnegieeurope.eu /2022/11/03/democratic-innovations-from-around-world-lessons-for-west-pub-88248.

Zacchiroli, Stefano. "Debian: 18 Years of Free Software, Do-ocracy, and Democracy." In *Proceedings of the 2011 Workshop on Open Source and Design of Communication—OSDOC '11*, 87. Lisbon: ACM Press, 2011. https://doi.org/10.1145/2016716.2016740.

Zargham, Michael. "Sensor Networks and Social Choice." BlockScience, 2019. https:// github.com/BlockScience/conviction/blob/master/social-sensorfusion.pdf.

Zargham, Michael, and Kelsie Nabben. "Aligning 'Decentralized Autonomous Organization' to Precedents in Cybernetics," 2022. SSRN, https://papers.ssrn.com/sol3/papers .cfm?abstract_id=4077358.

Zhang, Amy X., Grant Hugh, and Michael S. Bernstein. "PolicyKit: Building Governance in Online Communities." In *Proceedings of the 33rd Annual ACM Symposium on User Interface Software and Technology*, 365–78. New York, NY, USA: Association for Computing Machinery, 2020. https://doi.org/10.1145/3379337.3415858.

Zobl, Elke, and Ricarda Drüeke. *Feminist Media: Participatory Spaces, Networks and Cultural Citizenship*. transcript Verlag, 2012. https://doi.org/10.14361/transcript.9783839421574.

Zuckerberg, Mark. "Building Global Community." Facebook, February 16, 2017. https://www.facebook.com/notes/mark-zuckerberg/building-global-community/10154544292806634.

———. "Mark Zuckerberg's Letter to Investors: 'The Hacker Way.'" *Wired*, February 1, 2012. https://www.wired.com/2012/02/zuck-letter/.

———. "A Privacy-Focused Vision for Social Networking." Facebook, March 6, 2019. https://www.facebook.com/notes/mark-zuckerberg/a-privacy-focused-vision-for-social-networking/10156700570096634/.

Zuckerman, Ethan. "Cute Cats to the Rescue? Participatory Media and Political Expression." In *From Voice to Influence: Understanding Citizenship in a Digital Age*, edited by Danielle Allen and Jennifer S. Light. Chicago: University of Chicago Press, 2015. https://dspace.mit.edu/handle/1721.1/78899.

———. "How Social Media Could Teach Us to Be Better Citizens." *Journal of e-Learning and Knowledge Society* 18, no. 3 (December 28, 2022): 36–41. https://doi.org/10.20368/1971-8829/1135818.

INDEX

Founded in 1893,
UNIVERSITY OF CALIFORNIA PRESS
publishes bold, progressive books and journals
on topics in the arts, humanities, social sciences,
and natural sciences—with a focus on social
justice issues—that inspire thought and action
among readers worldwide.

The UC PRESS FOUNDATION
raises funds to uphold the press's vital role
as an independent, nonprofit publisher, and
receives philanthropic support from a wide
range of individuals and institutions—and from
committed readers like you. To learn more, visit
ucpress.edu/supportus.

Printed in the USA
CPSIA information can be obtained
at www.ICGtesting.com
JSHW011539040324
58548JS00008B/287